GEL ELECTROPHORESIS

GEL ELECTROPHORESIS
ESSENTIAL DATA

D. Patel

Department of Biology, University of Essex, Colchester, UK

JOHN WILEY & SONS
Chichester • New York • Brisbane • Toronto • Singapore

Published in association with BIOS Scientific Publishers Limited

©BIOS Scientific Publishers Limited, 1994. Published by John Wiley & Sons Ltd, Baffins Lane, Chichester, West Sussex PO19 1UD, UK in association with BIOS Scientific Publishers Ltd, St Thomas House, Becket Street, Oxford OX1 1SJ, UK.

British Library Cataloguing in Publication Data
A catalogue record for this book is available from the British Library.

ISBN 0 471 94306 1

Library of Congress Cataloguing in Publication Data
Patel, D.
    Gel electrophoresis : essential data / D. Patel.
      p.   cm.—(Essential data series)
    Includes bibliographical references and index.
    ISBN 0 471 94306 1
    1. Gel electrophoresis—Laboratory manuals.   I. Title.
II. Series.
QP519.9.G42P38   1994
574.19′285—dc20   94–5796
                                                                      CIP

Typeset by Marksbury Typesetting Ltd, Midsomer Norton, Bath, UK
Printed and bound in UK by H. Charlesworth & Co. Ltd, Huddersfield, UK

CONTENTS

Contents

ix

ABBREVIATIONS

A_{540} — absorption at 540 nm

% acrylamide — polyacrylamide gel concentration expressed in terms of *total* monomer (i.e. acrylamide and crosslinker)

APS — ammonium persulfate

bisacrylamide — *N, N'*-methylene bisacrylamide

%C — g crosslinker per 100 g monomers

Caps — 3-(cyclohexylamino)-1-propanesulfonic acid

conc. — concentrated

concn. — concentration

DipF — diisopropylphosphofluoridate

DEAE — diethylaminoethyl

DMSO — dimethyl sulfoxide

DNA — deoxyribonucleic acid

DNase — deoxyribonuclease

d.p.m. — disintegrations per minute

DTT — dithiothreitol

EDTA — ethylenediamine tetraacetic acid

EEO — electroendosmosis

Epps — 4-(2-hydroxyethyl)-1-piperazine-propane-sulfonic acid

EtOH — ethanol

Hepes — *N*-2-hydroxyethylpiperazine-*N'*-2-ethane sulfonic acid

HRP — horseradish peroxidase

IEF — isoelectric focusing

IPG — immobilized pH gradient

LGT — low gelling temperature

LMT — low melting temperature

M — molarity

MeOH — methanol

Mes — 2-(*N*-morpholino)-ethanesulfonic acid

Mops — 3-(*N*-morpholino)-propanesulfonic acid

Mol. wt — molecular weight

M_r — relative molecular mass

NEPHGE	non-equilibrium pH gradient electrophoresis		Sol.	soluble
			$\%T$	g monomers per 100 ml
No.	number		TAE	Tris–acetate–EDTA
NP-40	Nonidet P-40		TBE	Tris–borate–EDTA
PAGE	polyacrylamide gel electrophoresis		TCA	trichloroacetic acid
PFGE	pulse field gel electrophoresis		TEMED	N, N, N', N'-tetramethyl ethylenediamine
PMSF	phenylmethanesulfonyl fluoride			
PPO	2,5-diphenyloxazole		Tris	tris(hydroxymethyl)amino methane
PVDF	polyvinyldifluoride		UV	ultraviolet
RNA	ribonucleic acid		vol.	volume
RNase	ribonuclease		v/v	volume/volume
RT	room temperature		w/v	weight/volume
SDS	sodium dodecyl sulfate			

PREFACE

A wide range of electrophoresis techniques have been developed for the separation and analysis of all types of macromolecule, especially proteins and nucleic acids. Electrophoresis has become an indispensable tool for the research of cell and molecular biologists as well as biochemists because it is a simple, rapid and highly sensitive technique for both preparative and analytical separations.

This data book seeks to provide all of the core information required for carrying out all types of electrophoresis of proteins, nucleic acids and polysac-charides. This book provides a comprehensive reference source for reagents for electrophoresis, recipes for gels and buffers, sample preparation and analysis of gels. In addition, an extensive troubleshooting guide is provided to help the reader solve potential difficulties that may be encountered in carrying out gel electrophoretic separations.

I hope that the reader will find this book to be instructive and useful in their work on a daily basis.

D. Patel

HEALTH HAZARDS OF GEL ELECTROPHORESIS

Many chemicals commonly used for gel electrophoresis are toxic while the status of others is unknown. The precautions required for handling the chemicals mentioned in this text should be understood. Acrylamide, bisacrylamide, diethylbarbituric acid, ethidium bromide and formamide are toxic. Methylmercuric acid is poisonous and slightly volatile. Acrylamide is a known potent neurotoxin. Great care must be taken when handling such reagents.

It is essential to avoid any protease and nuclease contamination; therefore, where possible, all solutions should be autoclaved. Solutions which cannot be autoclaved should be prepared in glass double-distilled, autoclaved water and then filtered through a Millipore filter.

Chapter 1 **INTRODUCTION**

Electrophoresis is a method whereby charged molecules in solution, usually proteins and nucleic acids, migrate in response to an electrical field. Their rate of migration through the electrical field depends on the strength of the field, on the net charge, size and shape of the molecules, and on the ionic strength, viscosity and temperature of the medium in which the molecules are moving. As an analytical tool, electrophoresis is simple, rapid and highly sensitive. It is used to investigate the properties of a single charged species and also as a separating technique.

Electrophoresis is usually carried out in gels formed in tubes, slabs or on a flat bed. In a tube gel unit, the gel is formed in a glass tube, while a slab gel is formed in a glass sandwich made of two flat glass plates separated by two spacer strips at the edges and clamped together to make a water-tight seal. Both tube and slab gels are mounted vertically. For agarose gels in a flat bed unit the gel is poured on to a horizontal surface and has no cover plate on it; this cannot be used for polyacrylamide gels.

In most electrophoresis units, the gel is mounted between two buffer chambers or reservoirs, that contain separate electrodes such that the only electrical connection between the two chambers is through the gel.

The most commonly used support matrices in which the sample is run are agarose and polyacrylamide. They allow selective separation of molecules and provide a hard record of the electrophoretic run because at the end of the run the matrix can be stained and used for scanning, autoradiography or storage.

This book does not seek to duplicate the detailed methods for fractionating and analyzing proteins and nucleic acids

1

that are offered in many manuals — the reader should consult these manuals for such information. Rather, this book is a reference source for chemicals for electrophoresis, types of apparatus, recipes for gels and buffers, sample preparation, and analyses of gels. Also included are details of fractionation of carbohydrates by gel electrophoresis. Furthermore, a list of major suppliers of chemicals and apparatus required for gel electrophoresis is given.

The tables in this book are not intended to be exhaustive and there are recipes and referenced techniques other than those listed. The manufacturers and suppliers are given in abbreviated form, and the suppliers list in Chapter 8 should be consulted for the complete names and addresses.

Chapter 2 **CHEMICALS FOR ELECTROPHORESIS**

1 Acrylamide and bisacrylamide

Both acrylamide and bisacrylamide monomers are highly toxic, either by skin absorption or by inhalation of monomer powder. The effects range from skin irritation to central nervous system damage. A face mask, safety goggles and disposable gloves must be worn when handling monomers. Gloves must also be worn when handling the monomers in liquid form. Mouth pipetting should be prohibited. The purity of most commercial sources of monomers is adequate for most electrophoretic separations and does not require further purification. Electrophoresis grade monomers should be purchased. The polyacrylamide should be of low electroendosmosis type. Due to their toxicity it is inadvisable to attempt purification of acrylamide and bisacrylamide. After polymerization, the resulting polyacrylamide gel is essentially non-toxic. However, it is still advisable to wear gloves as a small proportion of monomer may remain unpolymerized and fingerprints may interfere with subsequent staining of the gel. Store acrylamide solutions in dark bottles at room temperature. Fresh solutions should be prepared every few months.

2 Agarose

The agarose should be of the low to medium electro-endosmosis type (EEO; i.e. EEO < 2.0). At very low concentrations this requisite may have to be sacrificed somewhat in favor of higher mechanical strength. Several specialist agaroses offering properties such as low-gelling (LGT) and low-melting (LMT) temperatures are available.

3 Sodium dodecyl sulfate

Only highly purified grades of sodium dodecyl sulfate

Chemicals for Electrophoresis

(SDS) for electrophoresis should be used. Select and use SDS from one source only. The use of different grades or sources of SDS can lead to marked alterations in the resolution, banding pattern and apparent molecular masses of polypeptides.

4 Urea

The main problem with using urea is the accumulation of cyanate ions in stock solutions due to chemical isomerization. The cyanate reacts with amino groups to form stable carbamylated derivatives which modify the charge of the proteins. As a result of incomplete reaction, several artifactual species of protein are generated. Therefore, it is essential to use fresh urea solutions and, where appropriate, to buffer the solutions with Tris. The free amino groups of Tris neutralize the cyanate ions. Where possible, heating of urea-containing solutions should be avoided as cyanate ion formation is accelerated with increasing temperature.

5 Tris

It is essential that Tris buffers are prepared with Tris base. After Tris base has been dissolved in deionized water, the pH of the solution should be adjusted with HCl. If Tris·Cl or Trizma are used to prepare buffers, the concentration of salt will be too high and polypeptides will migrate anomalously through the gel, yielding extremely diffuse bands.

6 Ammonium persulfate and riboflavin

Both ammonium persulfate (APS) and riboflavin are required electrophoresis grade, and are commercially available at purity of at least 98%. Solutions should be freshly prepared every few days and kept at 4°C.

7 TEMED

Most commercial sources are at least 99% pure. The TEMED used should be a colorless liquid.

All other reagents must be of the highest purity available. The purity of some reagents can be questionable and it is recommended that reagents for electrophoretic applications are purchased from suppliers who have quality-tested the products for use in these techniques. If it is necessary to purify any of the above reagents, refer to protocols described in literature listed in the Further Reading section.

Chapter 3 **TYPES OF APPARATUS**

Apparatus for gel electrophoresis is available from many commercial sources, some of which are listed in *Table 1*; this table lists the types of apparatus by their generic names and not by make or product number as companies continue to improve or discontinue particular products.

The supplier catalogs should be consulted for details regarding dimensions and other information such as voltage output. The manufacturers and suppliers are given in abbreviated form; the suppliers list in Chapter 8 gives the complete names and addresses.

Table 1. Types of apparatus and their suppliers

Apparatus	Manufacturer / supplier
Horizontal gel electrophoresis unit	ANA, APP, BDH, BML, BRL, BSS, ECA, FIL, FSE, GBL, GRI, HSI, IBI, JSL, NBL, PMB, RSL, SCC, SPL, SSI, STG, USL, VAH
Vertical gel electrophoresis unit	ANA, APP, BDH, BMI, BRL, BSS, ECA, FIL, FSE, GBL, GRI, HSI, IBI, JSL, MPC, RSL, SPL, SCC, STG, USL
Tube gel electrophoresis unit	BML, BRL, ECA, GRI, HSI, MPC, RSL
Capillary electrophoresis system	BRL, DNC, GRI
Immunoelectrophoresis system	BRL, ECA, HLL
Nucleic acid sequencing gel unit	ANA, APP, BML, BRL, ECA, FIL, FSE, GBL, GRI, HSI, IBI, JSL, PMB, RSL, SCC, STG
Pulse field gel electrophoresis (PFGE) unit	HSI, PMB
Fluorophore-assisted carbohydrate electrophoresis kits	GKI
Power supply	ANA, APP, BDH, BML, BRL, BSS, ECA, FIL, FSE, GBL, GRI, HSI, IBI, JSL, MPC, NBL, PMB, SCC, SPL, SSI, STG, USL, VAH

Spare gel glass plates	ANA, BML, BRL, BSS, ECA, FIL, FSE, GBL, GRI, HSI, IBI, PMB, RSL, SCC, SPL, STG, USL
Spare gel casting trays	ANA, APP, BML, BRL, BSS, ECA, FIL, FSE, GBL, GRI, HSI, IBI, NBL, PMB, RSL, SCC, SPL, STG
IEF gel tubes	BRL, ECA, HSI, MPC, RSL, SCC
IEF tube gel adaptor	BRL, HSI, RSL, SCC
Immuno frames	BRL, ECA
Electrodes	BRL, FSE, HSI, PMB
Transfer-blotting apparatus	ANA, APP, BDH, BML, BRL, BSS, ECA, FSE, GBL, GRI, HSI, JSL, MPC, PMB, RSL, SCC, SPL, STG
Gel dryer	APP, BDH, BML, BRL, ECA, FIL, FSE, GRI, HSI, VAH
Gel visualization / photodocumentation system	ANA, FIL, FSE, GRI, HLL, IBI, SCC, UVP, VBL
UV lamps	ANA, APP, GRI, HLL, IBI, USL, VBL
Bench top darkroom	ANA, FIL, USL, VBL
Imager system	APP, BRL, GRI, MDI, MPC, UVP
UV-transillumination device	ANA, APP, BDH, FIL, FSE, GRI, HSI, PMB, SCC, STG, USL, UVP
UV-crosslinking device	ANA, APP, BRL, FIL, HSI, STG, USL, UVP
Autoradiography cassettes	AIP, APP, GRI, SCC
Intensifying screens	AIP, APP, GRI, IBI, SCC
Densitometer	BRL, ECA, HLL, HSI, MDI, PMB
Software for gel reading	APP, BML, BRL, GRI, HSI, IBI, PMB, STG, UVP, VBL

Types of Apparatus

Chapter 4 GEL ELECTROPHORESIS OF PROTEINS

1 Recipes for gels and buffers

1.1 One-dimensional polyacrylamide gel electrophoresis

The analysis of complex mixtures of proteins and the determination of the purity of protein fractions has been possible using polyacrylamide gel electrophoresis (PAGE). The many methods can be categorized broadly into three main types: SDS-denaturing gels; non-denaturing gels; and isoelectric focusing gels.

SDS-denaturing gels

In SDS separations, migration is determined not by intrinsic electrical charge of polypeptides but by the molecular weight. SDS is an anionic detergent which, by binding to a protein, confers a net negative charge to the protein. The proteins then have a mobility which is inversely proportional to their size. Caution must be exercised, however, as some proteins bind less SDS and so migrate anomalously; the accuracy of size estimates by SDS–PAGE is taken to be about 10%. The Weber and Osborn system [1], although not commonly used today, is a continuous system. The Laemmli system [2] is a discontinuous SDS system and is probably the most widely used electrophoretic system today. The discontinuous system is used for maximal resolution of protein bands because the proteins are stacked in a stacking gel before entering the smaller-pore resolving or separating gel. *Table 1* gives recipes for the continuous buffer system, while *Table 2* gives recipes for discontinuous systems that are likely to be useful for most types of separations.

Recipes of modifications to the basic techniques, such as the continuous buffer system for the separation of oligopeptides (*Table 3*), or the use of linear gradients as opposed to step gradients (in which gels of different concentration are layered one upon the other) (*Tables 4* and *5*) are also given.

Non-denaturing gels

Electrophoresis under native conditions is used in circumstances where the activity or structure of a protein or protein complex under study is to be maintained. Native electrophoresis techniques can only be applied to protein samples which are soluble and which will not precipitate or aggregate during electrophoresis. In contrast to SDS–PAGE, the migration of proteins in non-denaturing gels is dependent on the amino acid composition of the protein and the pH of the electrophoresis buffer. It is essential to select the correct buffer to ensure optimal separation of the protein of interest. Again it is advantageous to use a discontinuous system to obtain stacking of the proteins and thus maximize resolution of the protein bands. Buffers and recipes for gels for the non-denaturing continuous system are given in *Tables 6* and *7*. *Tables 8* and *9* list buffers and recipes for gels for the non-denaturing discontinuous buffer system.

Isoelectric focusing

Isoelectric focusing (IEF) is a method in which proteins are separated in a pH gradient according to their isoelectric points, that is, the pH at which the net charge of the protein molecule is zero. At this pH, the protein molecules have no electrophoretic mobility and will be concentrated or focused into narrow zones. The most popular method for generating pH gradients for IEF is the incorporation of low molecular weight amphoteric compounds, synthetic carrier ampholytes, into a polyacrylamide gel matrix. When an electric field is applied the ampholyte molecules 'migrate' to one or other of the electrodes depending on their net charge. *Figure 1* shows the ranges of ampholyte mixtures that are available from different sources. *Table 10* lists solutions for IEF electrodes, and *Table 11* lists the choice of polyacrylamide gel concentration for protein separation. *Table 12* gives recipes for IEF gels.

IEF using immobilized pH gradients (IPGs), an alternative to IEF using carrier ampholytes, is often preferred.

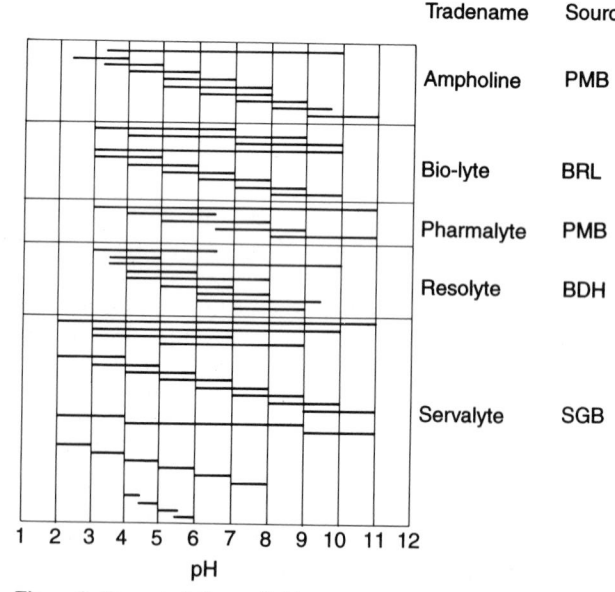

Tradename	Source
Ampholine	PMB
Bio-lyte	BRL
Pharmalyte	PMB
Resolyte	BDH
Servalyte	SGB

Figure 1. Commercially available carrier ampholytes.

IEF using IPGs is achieved by the incorporation of Immobiline reagents (Pharmacia Biosystems). The Immobiline reagents are a series of seven acrylamide derivatives, forming a series of buffers with different p*K* values distributed throughout the pH 3–10 range. *Table 13* lists the apparent p*K* values of the Immobiline chemicals. *Tables 14* and *15* list volumes of Immobiline required to prepare 1 pH or broad pH unit gradients, and *Table 16* gives recipes for forming two IPG gels. *Table 17* gives the concentration of Immobiline necessary to prepare 4, 5 and 6 pH unit gradients. Working concentrations for the catalysts for linear pH unit gradients and IPG electrode solutions are given in *Tables 18* and *19*, respectively.

Electrophoresis buffers (also termed electrode, reservoir, running, and tank buffers, and electrolytes) which facilitate passage of current and in turn the passage of proteins through the gel matrix from one electrode to the other, are also given.

1.2 Two-dimensional gel electrophoresis

Two-dimensional gel electrophoresis is widely used to separate complex mixtures of proteins into many more components than is possible in conventional one-dimensional electrophoresis. The polypeptides are separated on the basis of a different molecular property in each dimension. It is possible to resolve up to several thousand proteins on a single gel using two-dimensional separations. The most common two-dimensional electrophoresis method for analyzing polypeptides is to separate the proteins in the first dimension on the basis of charge by IEF and then to separate the polypeptides in the second dimension, in the presence of SDS, primarily on the basis of molecular mass of the polypeptides (*Tables 20* and *21*).

An alternative to IEF–SDS is non-equilibrium pH gradient electrophoresis (NEPHGE)–SDS. This system was developed by O'Farrell [3] for analyzing basic proteins. Most of the components, both chemical and those of the apparatus, are the same as in IEF–SDS. The major difference between the two systems is in the first dimension. Instead of applying the sample to the basic end of the gel, it is applied at the acidic end. All the proteins are thus positively charged and migrate towards the basic end of the gel. Care must be taken not to run the gels too long or the proteins will migrate too far (*Table 22*). Other systems that have been developed for particular types of proteins have also been described (*Tables 23–27*).

Also shown are electrophoresis buffers, equilibration buffers and sealing gels. Soaking in equilibration buffer leads to the exchange of ions present in the second-dimensional gel for contaminating ampholytes and electrolytes that are present from the first-dimensional gel and its electrophoresis. The sealing gel embeds the first-dimensional gel with the second-dimensional gel, such that there is a continuation between the two gels to facilitate complete migration of proteins from the first-dimensional gel into the second-dimensional gel during electrophoresis.

1.3 Immunoelectrophoresis

Immunoelectrophoresis is a procedure in which proteins and other antigenic substances are characterized by both their electrophoretic migration in a gel and their immunological properties. There are many variations of the technique but they are all based on the electrophoretic migration of antigens in an antibody-containing gel and specific immunoprecipitation of the antigens by means of corresponding precipitating antibodies. *Table 28* gives the reagents necessary for immunoelectrophoresis.

1.4 Gel electrophoresis using precast gels

Precast gels are available from many commercial sources. Instructions on rehydration are provided with the purchased gels and in literature listed in Further Reading. Some sources of precast gels are BRL, FIL, GRI, HLL, HSI, PMB and SCC (see Chapter 8 for full names and addresses).

2 Sample preparation

Table 29 summarizes procedures on solubilization of samples for analysis by gel electrophoresis. Non-ionic detergents such as Triton X-100 are particularly useful for solubilization of membranes and study of proteins, as they do not alter the properties of the proteins appreciably and have selectivity of solubilization. However, Triton X-100 and other non-ionic detergents disturb the patterns of SDS–PAGE. *Table 30* lists procedures for the removal of Triton X-100.

For small samples it is convenient to label the proteins to enhance detection after electrophoresis, reagents for isotopic labeling are given in *Table 31*. Sources of radioactive labels are AIP, ICN, NEN and SCC (see Chapter 8 for full names and addresses).

Ultimately, all proteins are degraded, either intracellularly or extracellularly; thus, under appropriate conditions, they can be completely hydrolyzed to amino acids by proteolytic enzymes. For this reason, inhibitors of proteases are often included in the samples. These are listed in *Table 32*. It must be remembered that some reagents can also modify proteins and introduce charge artifacts.

The tracking dyes used to follow migration of the samples/buffer front during electrophoresis are given in *Table 33*.

Samples are applied to the gel matrices using sample loading buffers (gel loading buffers) containing solubilizing or denaturing agents, inhibitors of proteases and tracking dyes as appropriate. Dense media such as sucrose or glycerol are also included to raise the density of the sample, and prevent the samples from floating out of the sample wells of the gel matrices and into neighboring wells or into the electrophoresis buffer. Solid samples can be solubilized directly in the sample loading buffer. Liquid or suspension samples should be diluted 1:1 with a double strength preparation of the sample loading buffer. *Table 34* gives sample loading buffers for various electrophoresis methods and applications.

In order to calibrate gels, whether they are SDS–PAGE or IEF gels, standard proteins are used as markers. *Tables 35* and *36* list standard marker proteins used in nondenaturing or denaturing conditions of electrophoresis. It is usual to select three or four proteins that migrate similarly to the protein of interest. Commercial sources of size markers are given in *Table 37*.

3 Analysis of gels

Proteins separated on gels can be visualized using a number of different procedures (*Tables 38–46*). Generally, staining using Coomassie blue is the most popular method, however if a more sensitive stain is required, silver staining is recommended.

If a protein or nucleic acid separation is transferred from a gel to a membrane such as nitrocellulose, it becomes extremely sensitive to detection techniques. All of the sample is bound to the surface of the membrane and is therefore available for the binding of specifically labeled probes or for autoradiography without quenching. *Table*

47 lists the many blotting matrices available. *Table 48* gives blocking solutions to prevent non-specific binding of probe to matrix. Transfer buffers for the transfer of protein to the blotting matrix are given in *Table 49*. There are many detection procedures for proteins on blots, and some of these are described in *Tables 50–55*. Included in the detection methods is ligand blotting. Proteins are detected or identified on the basis of their specific biological function. The proteins are fractionated by PAGE and transferred electrophoretically to a blotting matrix (e.g. nitrocellulose). Proteins are then identified on the matrix by their ability to bind a specific ligand. Fractionation by PAGE is achieved using either the gel system described by Laemmli [2] (*Table 2*) or that described in *Table 54*.

Suppliers of non-radioactive detection kits are given in *Table 56*. However, if the proteins are radioactive, they can be located by using autoradiography (*Tables 57–59*).

Renaturation of biological activity, after electrophoresis and subsequent blotting, is possible using methods given in *Table 60*. Renaturation may be required to gain further information about the protein structure and its identity or non-identity with other proteins.

Table 1. Recipe for gel preparation using the SDS–PAGE continuous buffer system [1]

Stock solution	% Acrylamide[a]				
	20.0	15.0	12.5	10.0	5.0
Acrylamide–bisacrylamide (30:0.8)	20.0	15.0	12.5	10.0	5.0
0.5 M sodium phosphate buffer pH 7.2	6.0	6.0	6.0	6.0	6.0
10% (w/v) SDS	0.3	0.3	0.3	0.3	0.3
Water	2.2	7.2	9.7	12.2	17.2
1.5% (w/v) APS	1.5	1.5	1.5	1.5	1.5
TEMED	0.015	0.015	0.015	0.015	0.015

Electrophoresis buffer: 0.1 M sodium phosphate pH 7.2, 0.1% (w/v) SDS.
[a]The columns represent volumes (ml) of stock solutions required to prepare 30 ml of gel mixture.

Table 2. Recipe for gel preparation using the SDS–PAGE discontinuous buffer system [2]

Stock solution	Stacking gel[a]	% Acrylamide in resolving gel[a]				
		20.0	15.0	12.5	10.0	5.0
Acrylamide–bisacrylamide (30:0.8)	2.5	20.0	15.0	12.5	10.0	5.0
0.5 M Tris–HCl pH 6.8	5.0	—	—	—	—	—
3.0 M Tris–HCl pH 8.8	—	3.75	3.75	3.75	3.75	3.75
10% (w/v) SDS	0.2	0.3	0.3	0.3	0.3	0.3
Water	11.3	4.45	9.45	11.95	14.45	19.45
1.5% (w/v) APS	1.0	1.5	1.5	1.5	1.5	1.5
TEMED	0.015	0.015	0.015	0.015	0.015	0.015

Electrophoresis buffer: 0.025 M Tris, 0.192 M glycine, pH 8.3, 0.1% (w/v) SDS.
Final concn. of buffers: stacking gel, 0.125 M Tris–HCl pH 6.8; resolving gel, 0.375 M Tris–HCl pH 8.8.
[a]The columns represent volumes (ml) of stock solutions required to prepare the gel mixtures.

Table 3. Recipe for separation of oligopeptides using urea–SDS–PAGE [4]

Gel mixture	12.5% acrylamide [acrylamide:bisacrylamide (10:1)], 0.1% (w/v) SDS, 8 M urea, 0.1 M H_3PO_4, Tris to final pH 6.8, 0.07% (w/v) APS, 0.075% (v/v) TEMED.

Electrophoresis buffer: 0.1% (w/v) SDS, 0.1 M H_3PO_4 adjusted to pH 6.8 with Tris base.

Table 4. Gel mixtures for a 5–20% SDS–PAGE gradient gel

Stock solution	% Acrylamide[a]	
	5.0	20.0[b]
Acrylamide–bisacrylamide (30:0.8)	5.0	20.0
3.0 M Tris–HCl pH 8.8	3.75	3.75
10% (w/v) SDS	0.3	0.3
Water	20.25	2.75
1.5% (w/v) APS	0.7	0.7
TEMED	0.010	0.010

Electrophoresis buffer: 0.025 M Tris, 0.192 M glycine, pH 8.3, 0.1% (w/v) SDS.
[a]The columns represent volumes (ml) of stock solutions required to prepare 30 ml of gel mixture.
[b]The 20.0% acrylamide mixture also requires the addition of 4.5 g sucrose (equivalent to 2.5 ml volume).

Table 5. Recipe for separation of low molecular mass polypeptides using gradient SDS–PAGE [5]

Stacking gel	5% (w/v) acrylamide, 0.13% (w/v) bisacrylamide, 0.067 M Tris–HCl pH 6.8, 0.1% (w/v) SDS, 0.067% (w/v) APS, 0.067% (v/v) TEMED.
Resolving gel	10–18% (w/v) linear acrylamide gradient, 0.5–0.9% (w/v) bisacrylamide, 0.45 M Tris–HCl pH 8.8, 0.1% (w/v) SDS, 7 M urea, 0–10% (w/v) linear gradient sucrose, 0.05% (w/v) APS, 0.05% (v/v) TEMED.

Electrophoresis buffer: 0.05 M Tris, 0.38 M glycine pH 8.5, 0.1% (w/v) SDS.

Table 6. Buffers for non-denaturing continuous buffer gel electrophoresis of proteins [6]

pH[a]	Acidic component	Basic component
3.8	20 mM lactic acid	30 mM β-alanine
4.4	40 mM acetic acid	80 mM β-alanine
4.8	20 mM acetic acid	80 mM γ-amino-n-butyric acid (Gaba)
6.1	30 mM Mes[b]	30 mM histidine
6.6	30 mM Mops[b]	25 mM histidine
7.4	35 mM Hepes[b]	43 mM imidazole
8.1	30 mM Epps[b]	32 mM Tris
8.7	25 mM boric acid	50 mM Tris
9.4	40 mM Caps[b]	60 mM Tris
10.2	20 mM Caps[b]	37 mM NH$_4$OH

[a]By varying the ratio of the basic and acidic components, pH values differing from that given by up to one pH unit may be obtained.
[b]See Abbreviations.

Table 7. Recipe for gel preparation using non-denaturing continuous buffer systems

	% Acrylamide[a]				
Stock solution	20.0	15.0	12.5	10.0	5.0
Acrylamide–bisacrylamide (30 : 0.8)	20.0	15.0	12.5	10.0	5.0
Continuous buffer (5 × concn.)[b]	6.0	6.0	6.0	6.0	6.0
Water	2.5	7.5	10.0	12.5	17.5
1.5% (w/v) APS	1.5	1.5	1.5	1.5	1.5
TEMED[c]	0.015	0.015	0.015	0.015	0.015

Electrophoresis buffer: for 1 liter, 200 ml continuous buffer (5 × concn.) and 800 ml water are required.
[a]The columns represent volumes (ml) of stock solutions required to prepare 30 ml gel mixture.
[b]See buffers, *Table 6*. Prepare five times the concentration described.
[c]The concn. of TEMED may need to be increased for low pH buffers.

Table 8. Buffers for non-denaturing discontinuous systems

Low pH discontinuous [7]:	stacks at pH 5.0, separates at pH 3.8
Stacking gel buffer	Acetic acid–KOH pH 6.8; 2.9 ml glacial acetic acid and 48.0 ml 1 M KOH mixed. Adjusted to final 100 ml volume.
Resolving gel buffer	Acetic acid–KOH pH 4.3; 17.2 ml glacial acetic acid and 48.0 ml 1 M KOH mixed. Adjusted to final 100 ml volume.
Electrophoresis buffer	Acetic acid–β-alanine pH 4.5; 8.0 ml glacial acetic acid and 31.2 g β-alanine dissolved in water. Adjusted to final 1 liter volume.
Neutral pH discontinuous [8]:	stacks at pH 7.0, separates at pH 8.0
Stacking gel buffer	Tris–phosphate pH 5.5; 4.95 g Tris dissolved in 40 ml water. Titrated to pH 5.5 with 1 M orthophosphoric acid. Adjusted to final 100 ml volume.
Resolving gel buffer	Tris–HCl pH 7.5; 6.85 g Tris dissolved in 40 ml water. Titrated to pH 7.5 with 1 M HCl. Adjusted to final 100 ml volume.
Electrophoresis buffer	Tris–diethylbarbiturate pH 7.0; 5.52 g diethylbarbituric acid and 10.0 g Tris dissolved in water. Adjusted to final 1 liter volume.
High pH discontinuous [9]:	stacks at pH 8.3, separates at pH 9.5
Stacking gel buffer	Tris–HCl pH 6.8; 6.0 g Tris dissolved in 40 ml water. Titrated to pH 6.8 with 1 M HCl (~48 ml). Adjusted to final 100 ml volume.
Resolving gel buffer	Tris–HCl pH 8.8; 36.3 g Tris and 48.0 ml 1 M HCl mixed. Titrated to pH 8.8 with HCl if needed. Adjusted to final 100 ml volume.
Electrophoresis buffer	Tris–glycine pH 8.3; 3.0 g Tris and 14.4 g glycine dissolved in water. Adjusted to final 1 liter volume.

Table 9. Recipe for gel preparation using non-denaturing discontinuous buffer systems

Stock solution	Stacking gel[a]	% Acrylamide in resolving gel[a]				
		20.0	15.0	12.5	10.0	5.0
Acrylamide–bisacrylamide (30:0.8)	2.5	20.0	15.0	12.5	10.0	5.0
Stacking gel buffer stock[b]	5.0	—	—	—	—	—
Resolving gel buffer stock[b]	—	3.75	3.75	3.75	3.75	3.75
Water	11.5	4.75	9.75	12.25	14.75	19.75
1.5% (w/v) APS	1.0	1.5	1.5	1.5	1.5	1.5
TEMED[c]	0.015	0.015	0.015	0.015	0.015	0.015

Electrophoresis buffer: as described in *Table 8*.

[a] The columns represent volumes (ml) of the various constituents required to prepare the gel mixtures.

[b] As described in *Table 8*.

[c] Increase volume of TEMED to 0.15 ml for the resolving gel when low pH discontinuous buffer system is used. Adjust volume of water accordingly.

Table 10. Electrolytes used in IEF

Solution	Concn. (M)	Use
CH_3COOH	0.5	Anolyte for alkaline pH ranges (pH > 7)[a]
H_2SO_4	0.1	Anolyte for very acidic pH ranges (pH < 4)[a]
H_3PO_4	1.0	Anolyte for all pH ranges
Histidine	0.2	Catholyte for acidic pH ranges (pH < 5)[b]
NaOH	1.0[c]	Catholyte for all pH ranges
Tris	0.5	Catholyte for acidic and neutral pH ranges (pH < 5)[b]

[a] Represents the lower limits of the pH range.

[b] Represents the higher limits of the pH range.

[c] Use air-tight plastic bottles for storage.

Table 11. Selection of polyacrylamide gel concentrations

Gel composition		Protein M_r ($\times 10^3$) (upper limit for a given %T)
%T	%C	
7	5	15
6	4	75
5	3	150
4	2.5	500

Table 12. Recipes for preparation of IEF gels (4–7% T and 4–8 M urea)

Gel vol. (ml)	30% T monomer solution[a] (ml)				2% Carrier ampholytes[b] (ml)		Urea (g)		TEMED (µl)	40% APS[c] (µl)
	4% T	5% T	6% T	7% T	A	B	4 M	8 M		
30	4.00	5.00	6.00	7.00	1.50	1.88	7.20	14.40	9.0	30
25	3.34	4.17	5.00	5.83	1.25	1.56	6.00	12.00	7.5	25
20	2.66	3.34	4.00	4.67	1.00	1.25	4.80	9.60	6.0	20
15	2.00	2.50	3.00	3.50	0.75	0.94	3.60	7.20	4.5	15
10	1.33	1.66	2.00	2.33	0.50	0.63	2.40	4.80	3.0	10
5	0.66	0.83	1.00	1.16	0.25	0.31	1.20	2.40	1.5	5

[a]Monomer solution:

30% T, 2.5% C; 29.25 g acrylamide and 0.75 g bisacrylamide mixed. Water added to final 100 ml volume.
30% T, 3.0% C; 29.10 g acrylamide and 0.90 g bisacrylamide mixed. Water added to final 100 ml volume.
30% T, 4.0% C; 28.80 g acrylamide and 1.20 g bisacrylamide mixed. Water added to final 100 ml volume.
[b]A, for 40% solution (Ampholine, Resolyte, Servalyte); B, for Pharmalyte.
[c]To be added after degassing the solution and just before pouring it into the gel mold.

Table 13. The apparent pK values of Immobiline chemicals in different media [10]

pK[a]	In water		In polyacrylamide gel 5%T, 3%C		In polyacrylamide gel 25% (w/v) glycerol 5%T, 3%C		Physical state at RT
	10°C	25°C	10°C	25°C	10°C	25°C	
Acids with carboxyl as the buffering group[b]							
3.6	3.57	3.58	—	—	3.68	3.75	solid
4.4	4.39	4.39	4.30	4.36	4.40	4.47	solid
4.6	4.60	4.61	4.51	4.61	4.61	4.71	solid
Bases with tertiary amines as the buffering group[c]							
6.2	6.41	6.23	6.21	6.15	6.32	6.24	solid
7.0	7.12	6.97	7.06	6.96	7.08	6.95	solid
8.5	8.96	8.53	8.50	8.38	8.66	8.45	liquid
9.3	9.64	9.28	9.59	9.31	9.57	9.30	liquid

[a]pK values measured with glass surface electrode without any correction at an ionic strength of 10^{-2}.

[b]Values given are mean pK data with average S.E. ± 0.02 in polyacrylamide gel ($\pm$ glycerol) .

[c]Values given are mean pK data with average S.E ± 0.06 in polyacrylamide gel ($\pm$ glycerol). Due to the slow response of the electrode the pK values are uncertain.

Table 14. Volume of Immobiline required to prepare 1 pH unit gradients [11]

pH range[a]	Control pH at 20°C	Acidic dense solution Volume (µl) 0.2 M Immobiline pK[b]							Control pH at 20°C	Basic light solution Volume (µl) 0.2 M Immobiline pK[b]						
		3.6	4.4	4.6	6.2	7.0	8.5	9.3		3.6	4.4	4.6	6.2	7.0	8.5	9.3
3.8–4.8	3.84±0.03	—	750	—	—	—	—	159	4.95±0.06	—	750	—	—	—	—	—
4.0–5.0	4.03±0.03	—	—	755	—	—	—	157	5.14±0.06	—	—	745	—	—	—	591
4.5–5.5	4.51±0.04	—	—	716	—	—	—	325	5.61±0.14	—	—	1314	—	—	—	1208
5.0–6.0	5.07±0.03	158	—	863	863	—	—	—	6.09±0.04	—	—	863	803	—	—	338
5.5–6.5	5.52±0.09	775	—	—	903	—	—	—	6.63±0.03	209	—	—	686	—	—	—
6.0–7.0	6.01±0.05	447	—	—	682	—	—	—	7.12±0.03	120	—	—	992	—	—	—
6.5–7.5	6.40±0.07	635	—	—	—	783	—	—	7.50±0.03	171	—	—	—	724	—	—
7.0–8.0	6.88±0.04	403	—	—	—	701	—	—	8.00±0.03	108	—	—	—	1217	—	—
7.5–8.5	7.66±0.15	1230	—	—	—	—	1334	—	8.76±0.04	331	—	—	—	—	720	—
8.0–9.0	8.13±0.06	591	—	—	—	—	750	—	9.24±0.06	159	—	—	—	—	750	—
8.5–9.5	8.62±0.04	389	—	—	—	—	720	—	9.74±0.06	105	—	—	—	—	1334	—
9.0–10.0	8.78±0.07	659	—	—	—	—	—	803	9.88±0.06	177	—	—	—	—	—	713
9.5–10.5	9.26±0.07	410	—	—	—	—	—	694	10.38±0.06	111	—	—	—	—	—	1165

[a]The pH range is that existing in the gel during the run at 10°C. For controlling the pH of the starting solutions, the values (control pH) are given at 20°C.

[b]Volumes of Immobiline for 15 ml of each starting solution.

Table 15. Volume of Immobiline required to prepare broad pH gradients [11]

pH range	Control pH at 20°C	Acidic dense solution Volume (µl) 0.2 M Immobiline pK^a						Control pH at 20°C	Basic light solution Volume (µl) 0.2 M Immobiline pK^a					
		3.6	4.6	6.2	7.0	8.5	9.3		3.6	4.6	6.2	7.0	8.5	9.3
3.5–5.0	3.53 ± 0.06	299	223	157	—	—	—	5.06 ± 0.07	212	310	465	—	—	—
4.0–6.0	4.00 ± 0.06	569	99	439	—	—	—	6.09 ± 0.14	390	521	276	—	—	722
4.0–7.0	4.01 ± 0.05	578	110	450	—	—	—	7.02 ± 0.14	302	738	151	269	—	876
5.0–7.0	5.08 ± 0.03	69	428	414	—	—	—	7.01 ± 0.06	—	474	270	219	—	320
5.0–8.0	5.03 ± 0.12	702	254	416	133	346	—	8.12 ± 0.07	175	123	131	345	346	—
6.0–8.0	6.06 ± 0.08	435	—	323	208	44	—	8.11 ± 0.09	286	—	174	325	329	—
6.0–9.0	6.04 ± 0.14	779	—	402	93	364	80	9.01 ± 0.06	241	—	161	449	237	225
7.0–9.0	7.03 ± 0.24	1349	—	—	272	372	845	8.94 ± 0.07	484	—	—	232	189	546
7.0–10.0	6.98 ± 0.07	542	—	—	378	351	—	9.88 ± 0.06	90	—	—	324	237	225
8.0–10.0	8.10 ± 0.07	399	—	—	364	355	94	9.89 ± 0.05	91	—	—	329	366	289

[a]Volumes of Immobiline for 15 ml of each starting solution.

Table 16. Recipe for the preparation of IPG gels [12]

Stock solutions	Acidic dense solution	Basic light solution
0.2 M Immobiline	—[a]	—[a]
Water	to 7.5 ml	to 7.5 ml
1M acetic acid[b]	—	titrate to pH 6.8
1M NaOH[b]	titrate to pH 6.8	—
Acrylamide–bisacrylamide (28.8:1.2)[c]	2 ml	2 ml
100% glycerol	3.8 g	—
Water	to 15 ml	to 15 ml
TEMED	10 μl	10 μl
40% (w/v) APS	16 μl	16 μl

[a]Volumes (μl) of 0.2 M Immobiline as given in *Tables 14* and *15*.
[b]Neutralization is not necessary when the mixture contains ampholytes in a pI range (e.g. 4–9.5) exhibiting a neutral pH prior to electrofocusing.
[c]These concentrations provide a 4%T, 4%C gel. Polyacrylamide stock solution must be modified to obtain alternative gel concentrations.

Table 17. Immobiline concentrations for 4, 5 and 6 pH unit gradients [13]

pH range	Control pH at 20°C	Acidic dense solution Concn. (mM) of Immobiline pK						Control pH at 20°C	Basic light solution Concn. (mM) of Immobiline pK					
		3.6	4.6	6.2	7.0	8.5	9.6		3.6	4.6	6.2	7.0	8.5	9.6
4.0–8.0	4.09	7.85	3.38	3.14	1.56	2.26	0	8.01	0	7.38	4.79	1.89	4.45	3.84
4.0–9.0	4.14	11.05	3.14	3.09	0.30	3.34	2.95	8.92	1.96	5.66	4.80	3.95	0.94	8.83
4.0–10.0	4.15	14.70	0	6.07	1.19	4.45	0	9.95	0	1.52	0.67	6.50	2.09	4.76
5.0–9.0	5.06	11.06	7.76	2.90	1.84	10.60	1.63	9.04	0	3.32	3.51	2.83	3.89	3.07
5.0–10.0	5.04	7.51	6.18	3.97	3.65	3.03	1.69	10.04	0.29	0.79	0.45	5.59	4.14	3.64
6.0–10.0	5.98	12.55	0	3.63	3.24	3.47	3.76	10.00	1.33	0	4.45	4.81	3.19	4.34

Table 18. Working concentrations of the catalysts for linear pH gradient gels

	Catalyst concn. (μl ml^{-1})		
	APS (40% w/v)	TEMED	
Conc. rating		Acidic pH	Basic pH
Lower limit	0.6	0.5	0.3
Standard, 3% T	1.0	0.7	0.5
Standard, 5% T [14]	0.8	0.5	0.3
For 5–10% alcohol	1.0	0.7	0.5
Higher limit [15]	1.4	0.9	0.6

Table 19. Electrolytes used in IPG

Solution	Concn.	Use
Carrier ampholytes[a,b]	0.3–1.0%	Anolyte and catholyte
Distilled water[b]	—	Anolyte and catholyte
Glutamic acid	10 mM	Anolyte
Lysine	10 mM	Catholyte

[a]Of the same or of a narrower range than the IPG.
[b]For mixed-bed gels or for samples with high salt concn.

Table 20. Recipes for first-dimensional IEF gel and second-dimensional SDS–PAGE gels [16]

First dimension

Gel mixture   1.33 ml 28.38% acrylamide, 1.62% bisacrylamide; 5.5 g ultrapure urea; 2 ml 10% NP-40; 0.4 ml 40% Ampholines (pH 5–7); 0.1 ml 40% Ampholines (pH 3.5 –10); 1.95 ml water; 10 μl 10% (w/v) APS; and 5 μl TEMED.
Electrolytes: anolyte, 10 mM H_3PO_4; catholyte, 20 mM NaOH.
Equilibration buffer: 2.5% (w/v) SDS, 5 mM DTT, 125 mM Tris–HCl pH 6.8, 10% (w/v) glycerol, 0.05% bromophenol blue.

Second dimension

Stock solution	Stacking gel[a]	Resolving gel	
		Light solution[a]	Dense solution[a]
29.2% acrylamide, 0.8% bisacrylamide	0.75	5.3	4.3
0.5 M Tris–HCl pH 6.8, 0.4% (w/v) SDS	1.25	—	—
1.5 M Tris-HCl pH 8.8, 0.4% (w/v) SDS	—	4.0	2.0
75% glycerol	—	—	1.7
Water	3.0	6.7	—
10% (w/v) APS	0.015	0.025	0.010
TEMED	0.005	0.008	0.004

Electrophoresis buffer: 25 mM Tris, 0.192 M glycine, 0.1% (w/v) SDS.
Sealing gel: 0.1% (w/v) agarose in 0.125 M Tris–HCl pH 6.8. Melt the agarose in Tris buffer and allow to cool to 55°C for normal agarose, or 45°C LGT agarose, before adding a tenth volume 20% (w/v) SDS.

[a]Columns represent volumes (ml) of stock solutions required to prepare the gel mixtures.

Table 21. Recipe for first-dimensional gels when the sample is mixed with the gels [17, 18]

Gel mixture:	0.5 ml sample in 8 M urea, 0.1% NP-40, 1 mM DTT. 0.133 ml 28.38% acrylamide, 1.62% bisacrylamide; 0.3 g ultrapure urea; 0.2 ml 10% NP-40; 0.05 ml 40% carrier ampholyte; 5 µl 10% (w/v) APS; and 2 µl TEMED.

Electrolytes: anolyte, (i) 10 mM H_3PO_4, (ii) 5% H_3PO_4, (iii) 100 mM Mops, 5% (w/v) sucrose; catholyte, (i) 20 mM NaOH, (ii) 5% 1, 2 diaminoethane, (iii) 100 mM Bis-Tris, 50% sucrose. Anolyte (i) used in conjunction with catholyte (i) and so on.

Table 22. Recipe for the first-dimensional gel for NEPHGE [3]

Gel mixture:	1.33 ml 28.38% acrylamide, 1.62% bisacrylamide (30% T, 5.7% C); 5.5 g ultrapure urea; 2.0 ml 10% NP-40; 0.5 ml 40% carrier ampholytes[a]; 1.93 ml water; 20 µl 10% (w/v) APS; and 14 µl TEMED.

The electrophoresis buffers and the second dimension are the same as that described in *Table 20*.
[a]Use Pharmacia Biosystems Ampholine 7–10 if basic proteins are to be analyzed. Use pH 3.5–10 if a wider range of pIs is to be studied.

Table 23. Two-dimensional electrophoresis of proteins under native conditions in the absence of urea or detergents [19]

	First dimension	Second dimension	
Stock solution	4.0%	4.0%	21.0%
Acrylamide–bisacrylamide (16:0.8)	4.0	10.0	—
Acrylamide–bisacrylamide (42:0.4) and 20% (w/v) sucrose	—	—	20.0
40% (w/v) Ampholine	0.8	—	—
0.3 M Tris–HCl pH 8.9, 0.23% TEMED	—	5.0	5.0
Water	1.2	15.0	10.0
0.1% (w/v) APS	8.0	10.0	5.0
0.23% (v/v) TEMED	2.0	—	—

Electrophoresis buffer: first dimension; anode, 0.01 M H_3PO_4, cathode, 0.04 M NaOH; second dimension; 0.05 M Tris–0.38 M glycine pH 8.3.

Table 24. Recipes for the separation of ribosomal proteins [20, 21]

Constituents	Gel mixtures[a]	
	First dimension pH 8.6	Second dimension pH 4.5
Acrylamide	6.0	18.0
Bisacrylamide	0.2	0.5
Boric acid	4.8	—
EDTA-Na$_2$	1.2	—
Tris-base	7.3	—
Urea	54.0	36.0
5 M KOH[b]	—	0.96
Glacial acetic acid[b]	—	5.3
TEMED[b]	0.45	0.58
Water[b]	to make 148.95 ml	to make 96.7 ml
10% (w/v) APS[b]	1.05	0.319

Electrophoresis buffer: 0.12 M Tris, 6 M urea, 6 mM EDTA-Na$_2$, 0.15 M boric acid, pH 8.6.

Tracking dyes: for first dimension, 0.5% pyronine G; for second dimension, 0.1% pyronine G in 20% glycerol.

Equilibration buffer: 8 M urea, 0.074% glacial acetic acid, 12 mM KOH, pH 5.2.

Sealing gel: second-dimensional resolving gel pH4.5.

[a]Columns represent mass (g) of constituents required unless otherwise indicated.

[b]Constituents required in volume (ml).

Gel Electrophoresis of Proteins

Table 25. Recipes for two-dimensional separation of histones on acid–urea gels

First-dimensional gel [22]

Stacking gel: 2.0 ml 10% acrylamide, 2.5% bisacrylamide; 1.0 ml 0.048 M KOH, 2.87% (v/v) glacial acetic acid, 0.46% (v/v) TEMED; 4.0 ml 9 M urea; 1.0 ml 0.4 mg ml^{-1} riboflavin.

Resolving gel: 4.0 ml 60% acrylamide, 0.4% bisacrylamide; 2.0 ml 0.48 M KOH, 17.2% (v/v) glacial acetic acid, 4% (v/v) TEMED; 9.0 ml 9 M urea; 50 μl 15% (w/v) APS.

Electrophoresis buffer: 3.12% β-alanine, 0.8% (v/v) glacial acetic acid. Pyronine G in urea used as marker dye solution.

Equilibration buffers: 1% (w/v) SDS, 1 mM DTT, 0.1 M phosphate, 10 mM Tris–HCl pH 7.0; 1% (w/v) SDS, 1 mM DTT, 0.01 M phosphate, 10 mM Tris–HCl pH 7.0; 0.1% (w/v) SDS, 1 mM DTT, 0.01 M phosphate, 10 mM Tris–HCl pH 7.0. Use each buffer in succession.

Second-dimensional gel [23]

Stacking gel: 3.33 ml 30% acrylamide, 0.8% bisacrylamide; 2.5 ml 1 M Tris–HCl pH 6.8; 0.2 ml 10% (w/v) SDS; 10 μl TEMED; water to 20 ml final volume; 100 μl 10% (w/v) APS.

Resolving gel: 30 ml 30% acrylamide, 0.15% bisacrylamide; 12.5 ml 3 M Tris–HCl pH 8.8; 0.5 ml 10% (w/v) SDS; 12.5 μl TEMED; water to 50 ml final volume; 125 μl 10% (w/v) APS.

Sealing gel: 1% (w/v) agarose in 0.0625 M Tris–HCl pH 6.8, 2.3% (w/v) SDS, 1 mM DTT.

Electrophoresis buffer: 0.05 M Tris base, 0.38 M glycine, 0.1% (w/v) SDS. Bromophenol blue used as marker dye.

Table 26. Recipes for two-dimensional separation of histones on Triton–acid–urea gels [24, 25]

Stock solutions	Gel dimension	
	First[a]	Second[a]
40% acrylamide, 0.267% bisacrylamide	3.0	15.0
43.2% (v/v) glacial acetic acid, 4% (v/v) TEMED	1.0	5.0
5 M urea, 0.2% (w/v) APS	4.0	—
2% (w/v) Triton X-100, 5 M urea, 0.2% (w/v) APS	—	20.0

Electrophoresis buffer: for first dimension use 0.9 M acetic acid; for second dimension use 0.9 M acetic acid, 1% (w/v) Triton X-100.
Methyl green used as tracking dye for first-dimensional separation.
[a]The columns represent volumes (ml) of stock solutions required for gel mixtures.

Table 27. Recipe for the two-dimensional separation of chromatin non-histone proteins [26]

First dimension: 5.0 ml 40% acrylamide, 1.4% bisacrylamide in 4 M urea; 5.0 ml 4% (v/v) TEMED in 2 M urea; and 5.0 ml 0.21% (w/v) APS, 21% (v/v) glacial acetic acid, in 6 M urea.
Second dimension: 18 ml 20% acrylamide, 0.52% bisacrylamide in 8 M urea; 7.5 ml 0.2% (v/v) TEMED, 0.4% (w/v) SDS in 0.4 M sodium phosphate pH 7.1; and 4.5 ml 0.5% (w/v) APS in 8 M urea.

Electrophoresis buffers: for first dimension, 0.9 M acetic acid in 4.5 M urea; for second dimension, 0.1 M sodium phosphate pH 7.1, 0.1% (w/v) SDS.
Equilibration buffers: 2% (w/v) SDS, 0.1 M sodium phosphate, 6 M urea, 1 mM DTT, pH 7.1; 1% (w/v) SDS, 0.01 M sodium phosphate, 6 M urea, 1 mM DTT, pH 7.1; 0.1% (w/v) SDS, 0.01 M sodium phosphate, 6 M urea, 1 mM DTT, pH 7.1. The equilibration buffers are used in succession.
Sealing gel: as described for second dimension gel except use 10 mM sodium phosphate pH 7.1.

Table 28. Buffer and gel reagents for immunoelectrophoresis

Gel mixture: 1% (w/v) agarose[a] in Tris–barbital buffer pH 8.6.
Tris–barbital buffer pH 8.6: 2.4 g 5,5'-diethyl barbituric acid; 44.3 g Tris; distilled water to final 1 liter volume. Dilute by a factor of 5 before use.

Electrophoresis buffer: Tris–barbital buffer pH 8.6.
Agarose is dissolved by boiling for 5 min. Kept molten in a water bath at 50–60°C. Once cooled and set, gel is stored at 4°C, and is ready for use once more after a short period of boiling.
[a]Use agarose with EEO $M_r = 0.13$ for normal procedures.

Table 29. Procedures and recipes used to solubilize samples for analysis by two-dimensional gel electrophoresis

Sample	Solubilizing buffer	Method	References
Bacterial proteins	9.5 M urea, 2% NP-40, 5% 2-mercaptoethanol	Freeze–thaw, sonication	16
Bacterial membranes	1. 50 mM Tris–HCl pH 6.8, 2% (w/v) SDS and 0.5 mM $MgCl_2$ 2. Add 2 volumes 9.5 M urea, 8% NP-40, 5% 2-mercaptoethanol, 2% Ampholines	Incubation at 70°C for 30 min	27
Fungal mycelium, membranes	1. 4 M guanidine thiocyanate, 1% 2-mercaptoethanol 2. Equilibrate sample with 8 M urea, 0.1% NP-40, 1 mM DTT	Blend using a suitable blender	28–30
Membranes, seed proteins	9.5 M urea, 5 mM K_2CO_3 pH 10.3	Sonication at low frequency	31
Nerve cells	1. 14 µl 1% (w/v) SDS, 10% 2-mercaptoethanol in 8 M urea 2. Add 4 mg solid urea, 4% Ampholines and 5 µl 10% NP-40	Homogenize in glass homogenizer	32
Animal cells	1. 3% (w/v) SDS, 10% 2-mercaptoethanol 2. Follow with incubation with DNase and RNase	Shear lyse cells and freeze, then lyophilize	33
Serum proteins	9 M urea, 0.1 M DTT	None as they are soluble proteins	28, 29

Table 30. Removal of Triton X-100 from protein samples prior to SDS–PAGE

Triton X-100 removal method	References
Chloroform extraction	34
Chromatography	35–37

Table 31. Reagents for the isotopic labeling of proteins

Reacting group	Labeling reagents
-CH_2OH. Aliphatic hydroxyl groups (serine/threonine residues)	Acetic anhydride, diisopropylphosphofluoridate (DFP)
-NH_2. Free amino groups (N-terminal or lysine residues)	Acetic anhydride, Bolton and Hunter reagent, dansyl chloride, ethyl acetimidate, 1-fluoro-2,4-dinitrobenzene, formaldehyde, isethionyl acetimidate, maleic anhydride, methyl 3,5-diiodohydroxybenzimidate, phenyl isothiocyanate, potassium borohydride, sodium borohydride, succinic anhydride, N-succinimidyl propionate
Imidazole groups (histidine residues)	Dansyl chloride, iodine
Phenolic hydroxyl groups (tyrosine residues)	Acetic acid, dansyl chloride, iodine
-SH. Thiol groups (cysteine residues)	Acetic acid, bromoacetic acid, chloroacetic acid, p-chloromercuribenzenesulphonic acid, p-chloromercuribenzoic acid, dansyl chloride, N-ethylmaleimide, iodoacetamide; iodoacetic acid

Table 32. Inhibitors of proteases

Inhibitor	Mol. wt	Effective concn.	Stock preparation	Additional information
Amastatin [(2S,3R)-3-amino-2-hydroxy-5-methyl-hexanoyl]-Val-Val-Asp-OH	474.6	1–10 μM	1 mM in methanol	Inhibitor of amino peptidases, in particular aminopeptidase N
4-Amidinophenyl-methanesulfonyl fluoride (APMSF, p-APMSF)	270.7 (APMSF·HCl·H$_2$O)	5–50 μM	20–50 mM in water, stable when aliquoted at −20°C	Specific for trypsin-like serine proteases. No effect on acetylcholine esterase. Not as effective as DipF, but more effective than PMSF
Antipain ([(S)-1-carboxy-2-phenylethyl]-carbamoyl-Arg-Val-Arg-al)	604.7	1–100 μM	10 mM in water or buffer; stable for 1 week at 4°C, 1 month at −20°C	Similar specificity to leupeptin, inhibits trypsin-like serine proteases and most cysteine proteases
Bestatin [(2S,3R)-3-amino-2-hydroxy-4-phenyl-butanoyl]-Leu-OH	308.4	1–10 μM	1 mM in methanol	Similar to specificity of amastatin
L-1-Chloro-3-[4-tosylamido]-7-amino-2-heptanone-HCl (TLCK, tosyl lysyl chloromethyl ketone, Tos-Lys-CH$_2$Cl)	369.4 (hydrochloride)	10–100 μM	10 mM in aqueous solution (pH 3–6); must be prepared fresh	Active towards some trypsin-like serine proteases; irreversible

L-1-Chloro-3-[4-tosylamido]-4-phenyl-2-butanone (TPCK, tosyl phenylalanyl chloromethyl ketone, Tos-Phe-CH$_2$Cl)	351.9	10–100 µM	10 mM in MeOH or EtOH; stable for several months at 4°C	Active towards some chymotrypsin-like serine proteases; irreversible
Chymostatin (Phe-(Cap)-Leu-Phe-al)	582.7	10–100 µM	10 mM in DMSO; stable for months at −20°C	Inhibits chymotrypsin-like serine proteases and most cysteine proteases; reversible
3,4-Dichloroisocoumarin (3,4-DCI)	215.0	5–100 µM	10 mM in dimethylformamide or dimethyl sulfoxide; stable at −20°C	Active towards wide range of serine proteases; slowly reversible; also active against acetylcholine esterase, kallikrein, trypsin, subtilisin, plasmin, thrombin, mast cell neutral protease, complement factor D, and Factors Xa, XIa, XIIa; not active towards β-lactamases
Diisopropylphosphofluoridate (diisopropylfluorophosphate, DipF, DFP)	184.2	0.1 mM	200–500 mM in dry propan-2-ol; stable for several months at −70°C	Active towards all serine proteases, irreversible, highly toxic
E-64 (L-trans-epoxysuccinyl leucylamido (4-guanidino)-butane)	357.4	10 µM	1 mM in aqueous solution; stable for months at −20°C	Effective irreversible inhibitor of cysteine proteases; does not affect cysteine residues in enzymes; active against papain and cathepsin B

Continued

33

Gel Electrophoresis of Proteins

Table 32. Inhibitors of proteases, *continued*

Inhibitor	Mol. wt	Effective concn.	Stock preparation	Additional information
EDTA	372.24 (di-sodium salt, di-hydrate)	1 mM	0.5 M in water, pH 8.5; stable for months at 4°C	EDTA acts as a chelator of the active site zinc ion in metallo-proteases; can also inhibit other metal ion-dependent proteases such as the calcium-dependent cysteine proteases; EDTA may interfere with other metal-dependent biological processes
Elastinal	512.6	10–100 μM	10 mM in water; stable for 1 week at 4°C, months at −20°C	Inhibits elastase-like serine proteases; reversible
Iodoacetate (IAA)	208.0 (sodium salt)	10–50 μM	10–100 mM in water; prepare fresh	Is not specific for the active site cysteine residue of serine proteases; can inhibit many other proteins and enzymes; active against thiol proteases
Leupeptin (*N*-acetyl-Leu-Leu-Arg-al)	542.7 (hemisulfate, monohydrate)	1–100 μM	10 mM in water or buffer; stable for 1 week at 4°C, 1 month at −20°C	Inhibits trypsin-like serine proteases and most cysteine proteases; reversible; also active against clostripain, cathepsin B and C

Pepstatin (Pepstatin A)	685.9	1 µM	1 mM in MeOH; stable for months at −20°C	Potent inhibitor of cathepsin D, pepsin, renin, acid proteases and many microbial aspartic proteases
1,10-Phenanthroline (orthophenanthroline)	198.2	1–10 mM	200 mM in MeOH; stable for months at −20°C	Inhibits metallo-proteases; has a strong UV absorbance
Phenylmethanesulfonyl fluoride (PMSF)	174.0	0.1–1 mM	20–50 mM in dry solvents (MeOH, EtOH, propan-2-ol); stable for at least 9 months at 4°C	Active towards all serine proteases; irreversible; not as effective or toxic as DipF, inhibits cysteine protease (reversible by reduced thiols); also active against chymotrypsin and trypsin
Phosphoramidon	543.6	1 µM	1 mM in water; stable for 1 month at −20°C	Inhibitor of few mammalian and many bacterial metallo-endopeptidases

Table 33. Tracking dyes for gel electrophoresis of proteins

Tracking dye	Application
Bromophenol blue	Neutral and alkaline tracking dye
Methyl green	Neutral tracking dye
Pyronine G	Two-dimensional gel electrophoresis

Gel Electrophoresis of Proteins

Table 34. Sample loading buffers for gel electrophoresis of proteins

Application	Sample loading buffer	Cross-references[a]
SDS–PAGE denaturing buffer system	2% (w/v) SDS, 5% (v/v) 2-mercaptoethanol, 10% (w/v) sucrose (or glycerol) and 0.002% bromophenol blue, in 0.01 M phosphate buffer pH 7.2 or 0.0625 M Tris–HCl pH 6.8	1, 2
Urea–SDS–PAGE for separation of oligopeptides	1% (w/v) SDS, 8 M urea, 1% 2-mercaptoethanol, 0.01 M H_3PO_4 adjusted to pH 6.8 with Tris base.	3
Gradient SDS–PAGE separation of low molecular mass polypeptides	0.0625 M Tris–HCl pH 6.8, 10% sucrose, 2% (w/v) SDS, 10 mM DTT or 1% 2-mercaptoethanol, 0.0025% bromophenol blue	5
Non-denaturing continuous buffer system	Use electrophoresis buffer, containing 10% (w/v) sucrose (or glycerol) and 0.002% tracking dye	7
Non-denaturing discontinuous buffer system	Use stacking gel buffer stock diluted 1/4–1/8, containing 10% (w/v) sucrose (or glycerol) and 0.002% tracking dye	9
First dimension IEF and NEPHGE	9.5 M urea, 5% 2-mercaptoethanol, 2% NP-40, 1.6% Ampholines (pH 5–7), and 0.4% Ampholines (pH 3.5–10)	20, 22
Two-dimensional electrophoresis of proteins under native conditions in the absence of urea or detergents	40% (w/v) sucrose	23
Two-dimensional separation of ribosomal proteins	50–100 μg protein in 0.1–0.15 ml with an equal volume of 1% (w/v) agarose in 0.12 M Tris, 6 M urea, 6 mM EDTA-Na$_2$, 0.15 M boric acid pH 8.6	24
Two-dimensional separation of histones on acid–urea gels	7 M urea	25

Two-dimensional separation of histones on Triton–acid–urea gels	8 M urea	26
Two-dimensional separation of chromatin non-histone proteins	0.9 M acetic acid, 10 M urea, 1% 2-mercaptoethanol	27
Non-denaturing Triton X-100 gradient gel electrophoresis for ligand blotting	20 mM Tris–glycine, pH 9.0, 2% (w/v) Triton X-100, 10% (w/v) glycerol	54
Fractionation of proteins on reduced gels containing SDS for ligand blotting	1.0 ml 0.5 M Tris–HCl pH 6.7, 0.46% (v/v) TEMED; 0.8 ml 10% (w/v) SDS; 0.8 ml glycerol; and 0.008 ml 2-mercaptoethanol	2[b]
Fractionation of proteins on non-reduced gels containing SDS for ligand blotting	1.0 ml 0.5 M Tris–HCl pH 6.7, 0.46% (v/v) TEMED; 0.8 ml 10% (w/v) SDS; 0.8 ml glycerol	2[b]

[a] The numbers listed in this column indicate the tables to which the reader should refer for further detail regarding the gel system used in conjunction with the sample loading buffers.
[b] Also refer to *Table 54*.

37   *Gel Electrophoresis of Proteins*

Table 35. Standard marker proteins used in non-denaturing gels [38, 39]

Polypeptide	Species	Tissue	Isoelectric point (pI)	No. of subunits	Subunit M_r ($\times 10^3$)
Adenine phosphoribosyl transferase	Human	Erythrocyte	4.8	3	11.0
Nerve growth factor	Mouse	Salivary gland	9.3	2	13.259
Ribonuclease	Bovine	Pancreas	7.8	1	13.7
Hemoglobin	Rabbit	Erythrocyte	7.0	4	16.0
Micrococcal nuclease	*S. aureus*	—	9.6	1	16.8
β-Lactoglobulin	Bovine	Serum	5.2	2	17.5
Ceramide trihexosidase	Human	Plasma	3.0	4	22.0
Adenylate kinase	Rat	Liver (cytosol)	7.5	3	23.0
Trypsinogen	Cow	Pancreas	9.3	1	24.5
Chymotrypsinogen A	Bovine	Pancreas	9.2	1	25.7
Triosephosphate isomerase	Rabbit	Muscle	6.8	2	26.5
Galactokinase	Human	Erythrocyte	5.7	2	27.0
Arginase	Human	Liver	9.2	4	30.0
Deoxyribonuclease I	Cow	Pancreas	4.8	1	31.0
Uricase	Pig	Liver	6.3	4	32.0
Glycerol-3-phosphate dehydrogenase	Rabbit	Kidney	6.4	2	34.0
Malate dehydrogenase	Pig	Heart	5.1	2	35.0
Alcohol dehydrogenase	Yeast	—	5.4	4	35.0
Deoxyribonuclease II	Pig	Spleen	10.2	1	38.0
Aldolase	Yeast	—	5.2	2	40.0
Pepsinogen	Pig	Stomach	3.7	1	41.0
Hexokinase	Yeast	—	5.3	2	51.0
Lipoxidase	Soybean	—	5.7	2	54.0

Catalase	Cow	Liver	5.4	4	57.5
Alkaline phosphatase	Calf	Intestine	4.4	2	69.0
Acetylcholinesterase	*Electrophorus*	—	4.5	4	70.0
Glyceraldehyde-3-phosphate dehydrogenase	Rabbit	Muscle	8.5	2	72.0
β-Glucuronidase	Rat	Liver	6.0	4	75.0
Lysine decarboxylase	*E. coli*	—	4.6	10	80.0
Glycogen synthetase	Pig	Kidney	4.8	4	92.0
Phosphoenolpyruvate carboxylase	*E. coli*	—	5.0	4	99.6
Phosphoenolpyruvate carboxylase	Spinach	Leaf	4.9	2	130.0
Urease	Jack bean	—	4.9	2	240.0

Gel Electrophoresis of Proteins

Table 36. Standard marker proteins used in denaturing gels [38, 40–42]

Polypeptide	Species	Tissue	M_r ($\times 10^3$)
Cytochrome c	—	—	11.7
Lysozyme	—	Egg white	14.3
α-Lactalbumin	Bovine	Milk	14.4
Myoglobin	Horse	Heart	16.95
Trypsin inhibitor	Soybean	—	20.1
Chymotrypsinogen A	—	—	25.7
Carbonic anhydrase	—	—	29.0
Lactate dehydrogenase	Pig	Heart	36.0
Glyceraldehyde-3-phosphate dehydrogenase	Rabbit	Muscle	36.0
RNA polymerase α-subunit	E. coli	—	39.0
Aldolase	Rabbit	Muscle	40.0
Alcohol dehydrogenase	Horse	Liver	41.0
Enolase	Rabbit	Muscle	42.0
Ovalbumin	—	—	43.0
Fumarase	Pig	Liver	48.5
Glutamate dehydrogenase	Bovine	Liver	53.0
Pyruvate kinase	Rabbit	Muscle	57.2
Catalase	Bovine	Liver	57.5
Bovine serum albumin	Bovine	Serum	68.0
Phosphorylase a	Rabbit	Muscle	92.5
β-Galactosidase	E. coli	—	130.0
RNA polymerase β'-subunit	—	—	155.0
RNA polymerase β'-subunit	E. coli	—	165.0
Myosin heavy chain	Rabbit	Muscle	212.0

Several proteases, such as trypsin, chymotrypsin, and papain, have been used as molecular mass standards but these may sometimes cause proteolysis of other polypeptide standards and thus are omitted here.

Table 37. Commercial sources of protein size markers[a]

Type of marker	Source
Protein markers	BDH, BHM, BRL, GBL, HSI, PMB, PML, SCC, STG
IEF markers	BRL, HSI, PMB, SCC
Two-dimensional markers	BRL, SCC

[a]Some of the suppliers offer prestained or radiolabeled size markers.

Table 38. Protein staining and detection methods used in IEF

Method	Application	Reference
Alcian blue	Glycoproteins	43
Autoradiography	Radioactive proteins	44
Blotting	Antigens	45
Coomassie blue G-250	General use	46
Coomassie blue G-250/urea/perchloric acid	In presence of detergents	47
Coomassie blue R-250/CuSO$_4$	General use	48
	In presence of detergents	48
Coomassie blue R-250/sulfosalicylic acid	General use	49
Copper stain	General use	50
Fast green FCF	General use	51
Fluorography	Radioactive proteins	52
Immunoprecipitation *in situ*	Antigens	53
Periodic acid–Schiff (PAS)	Glycoproteins	54
Print-immunofixation	Antigens	55
Silver stain	General use	56
Sudan black	Lipoproteins	57
Zymograms[a]	Enzymes	58

[a]The concentration of buffer in the assay medium usually needs to be increased to counteract the buffering action by carrier ampholytes.

Table 39. Staining procedures for two-dimensional polyacrylamide gels

Staining technique	Destaining technique	Comments
3–4 h in 0.1% Coomassie blue in methanol:water:acetic acid (5:5:1)	Overnight by diffusion against methanol:water:acetic acid (5:5:1)	—
20 min in 0.1% Coomassie blue in 50% TCA	Several changes of 7% (v/v) acetic acid	Removal of commercial ampholytes
3 h in 0.25% Coomassie blue in methanol:water:acetic acid (5:5:1)	Several changes of 5% (v/v) methanol, 10% (v/v) acetic acid	—
1–4 h in 0.1% Coomassie blue R-250 in 7.5% (v/v) acetic acid, 50% (v/v) methanol in water	Overnight in 7.5% (v/v) acetic acid, 50% (v/v) methanol in water	—
Overnight in 25% (v/v) isopropyl alcohol, 10% (v/v) acetic acid, 0.025–0.05% Coomassie blue, followed by 6–9 h in 10% (v/v) isopropyl alcohol, 10% (v/v) acetic acid, 0.0025–0.005% Coomassie blue	Several changes of 10% (v/v) acetic acid	An additional optional staining overnight in 10% (v/v) acetic acid containing 0.0025% Coomassie blue helps intensify the gel pattern
15 min in 0.55% amido black in 50% (v/v) acetic acid	40 h in 1% (v/v) acetic acid	—
3 h at 80°C, or overnight at RT in 0.1% amido black in 0.7% (v/v) acetic acid, 30% (v/v) ethanol in water	Several changes of 7% (v/v) acetic acid, 20% (v/v) ethanol in water	—
1 h in 50% (v/v) methanol, stain with fresh alkaline 0.8% AgNO$_3$ solution; wash with water for 5 min; to develop soak in fresh 0.02% HCHO, 0.005% citric acid for 10 min	Wash gel in water, transfer to 50% (v/v) methanol	Much more sensitive than Coomassie blue methods

Table 40. Staining procedures for proteins separated using immunoelectrophoresis

Stain	Staining procedure	Destaining procedure
Coomassie blue R-250	Stain gel for 5 min in 0.5% Coomassie blue R-250 in 96% EtOH:glacial acetic acid:water (4.5:1:4.5)	Destain 2–3 times, each for 5 min in 96% EtOH:glacial acetic acid:water (4.5:1:4.5)
Nigrosin	Stain gel until precipitates are stained sufficiently in 0.14% nigrosin in glacial acetic acid:0.1M sodium acetate:methanol: glycerol (2.15:5.75:1.4:0.7)	Destain in 20% (v/v) methanol in 5% (v/v) acetic acid followed by 5% (v/v) acetic acid and water

Table 41. Methods for staining proteins prior to electrophoresis

Stain	References
Remazol brilliant blue R	59
Drimarene brilliant blue K-BL	60
Serva blue G (synonymous with Coomassie brilliant blue G-250)	61

Table 42. *In situ* polypeptide detection methods in gels using fluorophore labeling

Fluorophore	References
Labeling with fluorophore prior to electrophoresis	
Dansyl chloride	62–64
Fluorescamine	65–68
DACM (*N*-(dimethylamino-4-methylcoumarinyl) maleimide)	69
MDPF (2-methoxy-2, 4-diphenyl-3(2H)-furanone)	67, 70, 71
o-Phthaldialdehyde	72
Labeling with fluorophore after electrophoresis	
Anilinonaphthalene sulfone (ANS)	73
Bis-ANS	74
Fluorescamine	75
p-Hydrazinoacridine	76
o-Phthaldialdehyde	77, 78

Table 43. *In situ* direct polypeptide detection methods in gels[a]

Detection method	Reference
Chilling	79
Precipitation with K^+ ions	80
Reaction with cationic surfactant	81
Sodium acetate	82
Via protein phosphorescence	83

[a]See also *Tables 39* and *42*.

Table 44. *In situ* methods for the detection of antigens in gels using immunological methods

Incubation of gel with	References
Antibody coupled to peroxidase followed by localization with 3,3′-diaminobenzidine	84–86
Fluorescein-labeled antibody	87, 88
Radiolabeled antibody	89, 90
Unlabeled antibody and then with ^{125}I–protein A	89, 91–93

Table 45. *In situ* detection of specific classes of proteins in gels

Method	References
Glycoproteins	
Crossed lectin electrophoresis	94
Fluorescent lectins	95
p-Hydrazino-acridine	96
Labeling of cell surface glycoproteins using galactose oxidase	97
Labeling of glycoproteins containing terminal *N*-acetylglucosamine using galactosyl transferase	98
Labeling glycoproteins *in vivo* using radiolabeled sugars	99–101
Lectins with covalently bound enzymes	102, 103
Periodic acid–Schiff (PAS)	95, 104–106
Periodic acid–silver stain	107
Radiolabeled lectins	89, 108, 109
Stains-all	110
Thymol–sulfuric acid	95
Lipoproteins	
Silver stain	111, 112
Staining before electrophoresis	113
Staining after electrophoresis	114
Phosphoproteins	
Entrapment of liberated phosphate (ELP)	115, 116
Silver stain	117
Stains-all	118
Trivalent metal chelation for acidic phosphoproteins (phosvitins)	116

Table 46. Detection of specific enzymes on gels

Enzyme	References	Enzyme	References
Acid phosphatase	119–122	Citrate synthase	121
Aconitase	120–122	Creatine kinase	119, 121, 122
Adenine phosphoribosyl transferase	121	3′ 5′ Cyclic AMP phosphodiesterase	119, 121
Adenosine deaminase	120, 121	Cysteine S-conjugate	125
Adenosine kinase	123	Cytidine deaminase	121
Adenylate kinase	120–122	DNase	126
ADP-glycogen transferase	119	DNA polymerase	119
Alanine aminotransferase	120, 121	Enolase	120, 121
Alcohol dehydrogenase	119–122	Esterases	119–122
Aldolase	120–122	Folate reductase	119
Alkaline phosphatase	119–122	β-D-Fructofuranosidase	120
Amine oxidase	119	Fructose-1,6-bisphosphate	122
Amino acid oxidase	119, 124	α-Fucosidase	119, 121
AMP deaminase	123	Fumarase	120–122
Amylase	119, 121	Galactokinase	119, 121
Arginase	123	Galactose-6-phosphate dehydrogenase	122
Arginosuccinase	123	Galactose-6-phosphate uridyltransferase	119, 121
Aromatic amino acid transaminase	122	α-Galactosidase	120, 121
Arylsulfatases	123	β-Galactosidase	119, 121
Aspartate aminotransferase	119–122	Glucanases	127, 128
Aspartate carbamoyl transferase	121	Glucose oxidase	119
Carbonic anhydrase	121, 122	Glucose-6-phosphate dehydrogenase	120–122
Cellobiose phosphorylase	119	Glucose phosphate isomerase	120
Cholinesterase	119	Glucose-1-phosphate uridylyltransferase	119, 121

Continued

Gel Electrophoresis of Proteins

Table 46. Detection of specific enzymes on gels, *continued*

Enzyme	References	Enzyme	References
Phosphoglyceromutase	120, 121	Sucrose phosphorylase	119
Phosphoglycollate phosphatase	123	Superoxide dismutase	121
Phosphatases	132	Thymidine kinase	122
Polynucleotide phosphorylase	119	Triose-phosphate isomerase	120–122
Pyridoxine kinase	124	UDPG dehydrogenase	119
Pyruvate kinase	120, 121	UDPG pyrophosphorylase	119, 121
Retinol dehydrogenase	122	UMP kinase	121
RNase	119, 122	Urease	119
Sorbitol dehydrogenase	121, 122	Xanthine dehydrogenase	122

Table 47. Blotting matrices

Blotting matrix	Additional information	Supplier[a]
Carboxy methyl membrane	For electrotransfer of small basic proteins. Ionic binding	S&S
Diazo-modified papers: diazobenzyloxymethyl (DBM) paper; diazophenylthioether (DPT) paper; cyanogen bromide-activated paper	DBM and DPT papers are not particularly stable and are prepared as amino-derivatives (ABM and APT) and diazotized prior to use. Resolution is generally less than with nitrocellulose. Have lower protein-binding capacities ($25–50$ μg cm^{-2}). Glycine, which is often a buffer component when electroblotting into nitrocellulose, cannot be used with DBM or DPT papers. However, since proteins are covalently bound, such papers may be the matrix of choice if a large number of detection methods are to be applied sequentially to the same blot. Cyanogen-bromide paper shares the limitations of the DBM and DPT papers	S&S
Ion-exchange paper, e.g. DEAE-paper	Useful for preparative work but the protein-binding capacity is low (~ 150 μg cm^{-2}). As proteins are bound only by charge interactions, care must be taken in the choice of transfer and processing solutions to avoid unacceptable losses of bound protein. Mechanically weak	S&S
Nitrocellulose membrane	Pure nitrocellulose membranes have good protein-binding capacity ($\sim 80–100$ μg cm^{-2}). Nitrocellulose membranes with 0.45 μm pore size are typically used but 0.22–0.1 μm has been recommended for lower molecular mass proteins. Recommended for immunodetection analyses due to reduced non-specific binding of antibodies, and for the analysis of basic proteins. Mixed ester membranes which contain cellulose acetate have reduced capacity	AIP, BML, BRL, EKL, GBL, HSI, MPC, PMB, RSL, S&S, SCC, STG

Continued

49

Table 47. Blotting matrices, *continued*

Blotting matrix	Additional information	Supplier
Nylon membrane	Far stronger and more robust than conventional pure nitrocellulose sheets. Have substantially higher protein-binding capacity, e.g. Zetaprobe (Bio-Rad) has approximately 6 times higher protein-binding capacity (~ 480 µg cm^{-2}) than nitrocellulose. Due to its highly cationic nature, recommended for electroblotting of SDS–PAGE gels to maximize binding of highly anionic SDS–polypeptide complexes. Charged nylon filters such as Zetaprobe recommended for electroelution of SDS–PAGE gels. Binding also much stronger than in the case of nitrocellulose. Uncharged nylon membranes should give higher binding of basic proteins. Major disadvantage of all nylon membranes is the lack of simple general staining procedures	AIP, BHM, BML, BRL, EKL, GBL, HSI, ICN, PMB, RSL, SCC, STG
Glass fiber	Recommended for amino and sequence analysis of separated proteins. Protein-binding capacity 10–20 µg cm^{-2}. The proteins can be blotted from SDS–PAGE gels and then acid-hydrolyzed directly while still immobilized on the filter	BML, JNP
Polyamide membranes	Hydrophilic in nature. Strong binding of nucleic acids by electrostatic forces. High binding capacity and very high sensitivity. Available either uniformly charged with equal amounts of positive (amino groups) and negative (carboxylic groups) charges or charged with strongly cationic groups (quaternary ammonium groups)	S&S
Polyvinyldifluoride (PVDF)	Hydrophobic in nature. Compatible with commonly used protein stains as well as standard immunodetection methods	BRL, ICN, MPC, S&S, SCC
Blotter/filter paper	Provide a uniform flow of buffer through the gel to the transfer membrane. Ensure using correct quality of paper as different blotting applications require different strength paper	BML, BRL, HSI, S&S, SCC

[a]See Chapter 8.

Table 48. Blocking solutions to prevent non-specific binding of the probe to the matrix in protein blotting [133]

Blocker	Concn. (% w/v)
Bovine serum albumin (BSA)	0.5–10
Casein	1–2
Ethanolamine	10
Fetal calf serum	10
Gelatin	0.25–3
Hemoglobin	0.1–3
Milk	5
Newborn calf serum	5
Ovalbumin	1–5
Polyvinylpyrrolidone	2
Tween 20	0.05–0.5

Table 49. Transfer buffers used in protein blotting

Transfer buffer	Additional information	References
25 mM Tris, 192 mM glycine, 20% methanol, pH 8.3	0.05–0.1% SDS can be included	45
48 mM Tris, 39 mM glycine, 20% methanol, pH 9.2	0.0375% SDS can be included	134
10 mM $NaHCO_3$, 3 mM $NaCO_3$, 20% methanol, pH 9.9		135
0.7% Acetic acid		136

Table 50. General protein stains for blot transfers

Protein stain	Blotting matrix	Sensitivity
Amido black 10B	Nitrocellulose, PVDF	1.5 μg
Colloidal gold	Nitrocellulose, PVDF	4 ng
Colloidal iron	Nitrocellulose, nylon, PVDF	30 ng
Coomassie brilliant blue R-250[a]	Nitrocellulose, PVDF	1.5 μg
Fast green FC	Nitrocellulose, PVDF	—
India ink	Nitrocellulose, PVDF	100 ng
In situ biotinylation + HPR-avidin	Nitrocellulose, nylon, PVDF	30 ng
Ponceau S	Nitrocellulose, PVDF	—
Silver-enhanced copper	Nitrocellulose, nylon	—

[a]Results obtained in high background.

Table 51. Specific detection methods for blot transfers

Detection method	Approximate sensitivity (ng mm^{-2})
Avidin–biotin–peroxidase (ABC) complex	0.5
Enhanced chemiluminescence	0.001
Gold–second antibody	1.5
Gold–second antibody with silver enhancement	0.1
^{125}I–second antibody	1.0
Peroxidase–antiperoxidase	0.5
Peroxidase double sandwich	0.8
Peroxidase–protein A	2.0
Peroxidase–second antibody	1.5

Table 52. Detection of proteins on blots

Proteins detected	References
Glycoproteins	103, 137, 138
Lipoproteins	139
Radioactive proteins	45, 140

Table 53. Immunological detection methods on blots

Detection method	References
Biotin-conjugated antibody	30
Enzyme-conjugated antibody	141, 142
Fluorescent-labeled antibody	45, 143
Immunogold staining	144
Labeled *S. aureus* protein A	30, 145
Radiolabeled antibody	30, 45

Table 54. Solutions for non-denaturing Triton X-100 gradient gel electrophoresis for ligand blotting [146]

Stock solution	Sample gel (3%)[a]	Light gel (3%)[a]	Heavy gel (20%)[a]	Heavy gel (15%)[a]
Acrylamide–bisacrylamide (30:0.8)	1.5	1.75	11.67	8.75
2 M Tris–glycine, pH 9.0	0.75	0.875	0.875	0.875
10% (w/v) Triton X-100	0.15	0.175	0.175	0.175
Water	12.25	14.3	—[b]	—[b]
1.5% (w/v) APS	0.35	0.41	0.41	0.41
TEMED	0.005	0.006	0.006	0.006

Electrophoresis buffer: 100 mM Tris base pH 9.0, 0.1% (w/v) Triton X-100.
[a]Columns represent volumes (ml) of stock solutions required to prepare gel mixtures.
[b]For heavy gel solution, 2.63 g sucrose is added and the volume adjusted to 17.5 ml.

Gel Electrophoresis of Proteins

Table 55. Staining procedures for gels and blots in ligand blotting

Staining technique	Destaining technique	Application
2 h or overnight in 0.2% (w/v) Coomassie blue R-250 in 12.5% (v/v) glacial acetic acid, 19% (v/v) methanol in water	Destain until background is clear in methanol:glacial acetic acid:water (4:1:6)	Protein stain for acrylamide gels
10 minutes in 0.1% (w/v) amido black in glacial acetic acid:methanol:water (1:4.5:4.5)	Destain until background is clear in 10% (w/v) glacial acetic acid	Protein stain for nitrocellulose membranes

Table 56. Suppliers of non-radioactive detection kits

Method of detection	Supplier
Chemiluminescence	BRL, MPC, NEN, PML, S&S, USB
Colloidal gold systems	BRL
Enzyme systems	BRL, PML, S&S

Table 57. *In situ* detection of radioactive proteins in gels[a]

Detection method	References
Double-label detection using X-ray film	147, 148
Electronic data capture	149, 150
Fluorography using PPO in DMSO	44, 151
Fluorography using PPO in glacial acetic acid	152
Fluorography using sodium salicylate	151, 153, 154
Fluorography using commercial reagents	140, 155
Image intensification	156
Indirect autoradiography using an X-ray intensifying screen	157, 158
Quenching of radiolabeled proteins by gel conditions	159
Radiolabeling proteins *in vivo* prior to electrophoresis	30, 160
Radiolabeling proteins *in vitro* prior to electrophoresis	30
Radiolabeling proteins after gel electrophoresis	161, 162

[a]See *Tables 44* and *45* for radiolabeled probes.

Table 58. Sensitivities of methods for radioisotope detection in polyacrylamide gels [163]

Isotope	Method	Detection limit d.p.m. cm^{-2} for 24 h	Relative performance compared to direct autoradiography
^3H	Direct autoradiography	$> 8 \times 10^6$	1
	Fluorography using PPO	8000	> 1000
^{14}C or ^{35}S	Direct autoradiography	6000	1
	Fluorography using PPO	400	15
^{32}P	Direct autoradiography	525	1
	Intensifying screen	50	10.5
^{125}I	Direct autoradiography	1600	1
	Intensifying screen	100	16

The data are for exposure at $-70°C$ using X-ray film pre-exposed to $A_{540} = 0.15$ above the background absorbance of unexposed film.

Table 59. Films for autoradiography

Film	Source
AGFA	APP
Fuji-AX	GRI
GRI-AX	GRI
Hyperfilm	AIP
Hyperpaper	AIP
Kodak X-OMAT AR	EKL, IBI, SCC

Table 60. Procedures for the restoration of biological activity after gel electrophoresis

Renaturation technique	References
In situ	108, 164–166
After electroblotting	167, 168
After elution into free solution	169–171

Chapter 5 GEL ELECTROPHORESIS OF NUCLEIC ACIDS

1 Recipes for gels and buffers

1.1 RNA

Nucleic acids can be separated on gel matrices on the basis of their size and conformation. The gel matrices used are either polyacrylamide or agarose gels. Polyacrylamide gels are usually used for separating smaller molecules. Gel electrophoresis of RNA can be carried out under non-denaturing or denaturing conditions. In the former case, the secondary structure of the RNA is maintained and the migration of the RNA molecules is not directly related to their molecular weights. Alternatively, under denaturing conditions, hydrogen bonding is disrupted and the RNA migrates as single-stranded molecules. The rate of migration of RNA during electrophoresis under denaturing conditions can provide accurate determinations of molecular weights. Buffers used for gel electrophoresis of RNA are listed in *Table 1*. Recipes for non-denaturing

polyacrylamide tube gels are given in *Table 2*. *Tables 3–5* give recipes for non-denaturing slab gels. Some denaturants used in gel electrophoresis of nucleic acids (RNA and DNA) are listed in *Table 6*. *Tables 7–9* give recipes for denaturing polyacrylamide gels, whereby the denaturants, formamide, urea, and methylmercuric acid are added to the gel mixture. Agarose gels are very simple to prepare: the agarose is heated in the buffer of choice to obtain a molten gel (a microwave can be used for this), poured and then allowed to cool. Recipes for non-denaturing and denaturing agarose gels are given in *Table 10*. However the method most frequently used for fractionating RNA is by electrophoresis in horizontal agarose slab gels, following denaturation of the RNA at high temperatures in the presence of glyoxal, dimethylsulfoxide, formaldehyde and formamide. Recipes for fractionation after denaturation of RNA using the aforementioned denaturants are given in *Tables 11* and *12*.

57

1.2 DNA

Electrophoresis of DNA may be carried out in agarose or polyacrylamide gels. The most common application is the separation of duplex DNA fragments at neutral pH values. Agarose gels can be used to analyze double-stranded DNA fragments from 70 base pairs (3% agarose gel) to 800 000 base pairs (0.1% agarose gel). Polyacrylamide gels are used for fragments between 6 base pairs (20% acrylamide) and 1000 base pairs (3% acrylamide). Analysis of single-stranded DNA fragments is possible using the several types of denaturing gel systems available today. The fragments migrate with a size-dependent mobility. Single strands of DNA moving in non-denaturing gels show some dependence of mobility on secondary structure. Buffers and recipes for non-denaturing polyacrylamide gels are given in *Tables 13* and *14*. *Tables 15–17* give buffers and recipes for denaturing polyacrylamide gels. Recipes for agarose gels for the separation of DNA fragments are given in *Table 18*.

1.3 Pulse field gel electrophoresis (PFGE)

New developments in DNA electrophoretic separation have enabled the resolution of DNA molecules in size ranges greater than 10 megabases. It has been found that by periodically changing the direction of the migration of a mixture of large DNAs, the molecules separate from each other according to size. The size of the molecules that will separate from each other depends on the length of the pulse in each direction. A fast pulse rate will cause small molecules to separate, while a slow pulse rate separates large molecules. The different electrode configurations used in the different types of PFGE apparatus are shown in *Figure 1*. Electrophoresis buffers for PFGE are given in *Table 19*.

1.4 Synthetic oligonucleotides

The electrophoresis of synthetic oligonucleotides is essentially a subset of the techniques used to sequence DNA. The basic techniques, equipment and buffer systems are the same. Synthetic oligonucleotides can range in size from several nucleotides to 200 nucleotides in length. Denaturing polyacrylamide gels provide separations based on the molecular weight assuming a constant ratio of charge to mass (*Table 20*).

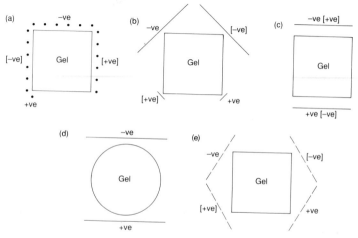

Figure 1. Electrode configurations used in the different types of PFGE apparatus. Changes in polarities after the duration of a pulse are shown in brackets. (a) Diode isolated electrodes (•) providing one homogeneous and one inhomogeneous field; (b) double inhomogeneous field using long cathodes and short anodes; (c) field inversion gel electrophoresis using a homogeneous electrical field; (d) rotating gel system using a homogeneous field (the polarity remains the same but the gel is turned at each pulse interval); (e) contour clamped homogeneous electric field system uses a hexagonal electrophoresis chamber with electrodes on four sides of the hexagon. Reproduced from R. Anand and E.M. Southern (1990) [in *Gel Electrophoresis of Nucleic Acids, a Practical Approach*, 2nd Edn. IRL Press, Oxford] with permission from Oxford University Press.

1.5 Two-dimensional gel electrophoresis

Many two-dimensional systems have been described for the separation of nucleic acids too complex for fractionation in a single dimension. Examples of procedures for fractionation of oligonucleotides (*Tables 21* and *22*), RNA (*Tables 23–25*) and DNA (*Tables 26* and *27*) are given. Also shown are electrophoresis and equilibration buffers, and sealing gels.

1.6 Nucleic acid–protein interactions

Simple electrophoretic methods for detecting nucleic acid–protein interactions are described (*Tables 28* and *29*). Gel retardation analysis allows the separation and quantification of free and complexed components of a binding reaction. In an acrylamide gel, mobility of a specific DNA–protein complex may differ substantially from that of the free DNA. During electrophoresis free DNA will rapidly enter the gel as a band, physically removed from the other components. The highly negatively charged DNA will pull bound proteins into the gel, but DNA–protein complexes are retarded in moving through the gel matrix.

1.7 DNA sequences

In order to purify and characterize sequence-specific DNA-binding proteins, techniques have been developed to detect their presence in cell extracts. *Tables 30–33* list recipes for the analyses of the resulting DNA fragments containing the sequences of interest.

1.8 Nucleosomes

Nucleosomes are made up of both protein and DNA. They lend themselves to analysis by two-dimensional electrophoresis. Nucleosomes are separated by electrophoresis in a 5% polyacrylamide gel at low ionic strength in the first dimension. The protein is then dissociated from the DNA and the DNA and proteins are analyzed in two different second-dimensional gels. Recipes for one-dimensional electrophoresis of nucleosomes are described in *Tables 34* and *35*. These gels also serve as first-dimensional gel electrophoresis systems for two-dimensional electrophoretic analyses; recipes for the preparation of the second-dimensional gel are given (*Tables 36* and *37*).

1.9 Polysomes and ribosomes

The preparation of agarose–polyacrylamide composite gels for the electrophoretic separation of polysomes and ribosomes are given in *Table 38*. In the case of nucleoproteins, gels of a low concentration (approx. 2%) of polyacrylamide are strengthened by the addition of a low concentration (0.5%) of agarose.

1.10 Gel electrophoresis using precast gels

Precast gels for gel electrophoresis of nucleic acids are available from a number of commercial sources: BRL, FIL and SSC (see Chapter 8 for full names and addresses). Fewer suppliers are available in comparison to those for gel electrophoresis of proteins. Instructions on rehydration are provided with the purchased gels and in literature listed in Further Reading.

2 Sample preparation

Non-radioactive and radioactive labeling methods enhance detection of nucleic acids after electrophoresis.

Examples of such methods are given (*Tables 39* and *40*). Sources of radioactive labels are listed in *Table 41*.

As with proteins, nucleic acids are susceptible to degradation. In this case degradation is effected by nucleases which are subdivided into either deoxyribonucleases (DNases) or ribonucleases (RNases). DNases can be inhibited by chelating agents if metal-dependent, by changing the pH or ionic strength outside the working range, by low concentration of denaturants, or by gentle heating. Many nucleic acid binding antibiotics (e.g. duanomycin, ethidium bromide) also inhibit DNases. RNases can be inhibited by a number of inhibitors. Metal-dependent enzymes may be inhibited by chelating agents. Many of the chemical inhibitors of RNases bind to the nucleic acids and prevent the enzyme from acting by steric effects. It is advisable to use high concentrations of strong denaturing agents such as SDS, guanidinium salts, 6 M urea, or 3 M LiCl during the isolation of RNA. Selected inhibitors of nucleases are given in *Table 42*.

Tracking dyes used to follow the migration of nucleic

acids/buffer front during electrophoresis through either agarose or polyacrylamide gels are listed in *Table 43* and sample loading buffers used for preparing and loading the sample on to the gels are described in *Table 44*.

Calibration of the gels is achieved by running nucleic acid size markers alongside the sample in the same gel. By comparison of the migration of size markers and that of samples, the sizes of the nucleic acid sample components can be determined. A number of naturally occurring species of RNA can be used as markers, and a selection of these is listed in *Table 45*. DNA markers can be prepared from restriction nuclease digests of plasmid or viral DNA. A selection of the restriction digests of three such digests are given (*Tables 46–48*). Suppliers of nucleic acid size markers are listed (*Table 49*).

3 Analysis of gels

Nucleic acids separated on gels can be visualized *in situ* using a number of different staining procedures (*Table 50*). Alternatively, the nucleic acids can be transferred to blotting matrices on which subsequent detection can proceed. (Blotting matrices and their properties are listed in *Table 47* of Chapter 4.) Denaturation of DNA in gels for transfer to nitrocellulose membrane is required (*Table 51*). Denaturation is required to prevent re-association of single-stranded DNA during transfer to nitrocellulose membranes. Alkaline conditions prevent re-association during transfer. Transfer to nitrocellulose in alkaline conditions is not possible because nitrocellulose disintegrates in alkali.

Transfer buffers for the transfer of nucleic acids to blotting matrices are given in *Table 52*. Blocking (or prehybridization) and hybridization solutions used to probe the blotting matrices are given in *Table 53*. Different nucleic acid strands are annealed during hybridization, to probe for particular gene sequences.

If the nucleic acids have been radioactively labeled, they can be located using autoradiography. There are a number of isotope detection methods available (*Table 54*). Preparation for fluorography is described (*Table 55*).

(Sensitivities of methods for radioisotope detection in polyacrylamide gels and films for autoradiography are listed in *Tables 58* and *59* of Chapter 4.)

The nucleic acids may be recovered directly from the gel matrices after electrophoresis for further analysis (*Table 56*).

Table 1. Electrophoresis buffers for RNA separation

Buffer	Final concn.	Additional information
1. Tris base EDTA·Na$_2$·2H$_2$O SDS HCl	30 mM 0.1 mM 0.1% 16 mM	The buffer should be pH 8.0 at RT; changes in conformation or mobility may be observed if 2 mM magnesium acetate is added to this buffer; the gels are electrophoresed at 25°C to prevent the SDS precipitating
2. Tris base NaH$_2$PO$_4$·2H$_2$O EDTA·Na$_2$·2H$_2$O SDS	3.6 mM 3 mM 0.1 mM 0.1%	The buffer should be pH 7.7 at RT; it is best to circulate the solution between the two buffer reservoirs during electrophoresis, to minimize pH changes
3. Tris base Boric acid EDTA·Na$_2$·2H$_2$O SDS	90 mM 90 mM 2.5 mM 0.1%	The buffer should be pH 8.3 at RT
4. Tris base NaH$_2$PO$_4$·2H$_2$O EDTA·Na$_2$·2H$_2$O SDS	36 mM 30 mM 1 mM 0.2%	The buffer should be pH 7.7 at RT

The corresponding buffer used to prepare the gels should be five times the above concentrations and without SDS.

Table 2. Recipes for non-denaturing polyacrylamide tube gels for RNA separation

	% Acrylamide[a]						
Stock solution	10.0	7.7	5.0	4.0	3.0	2.6	2.2
Acrylamide–bisacrylamide (15:0.75)[b]	10.0	10.0	10.0	5.0	5.0	5.0	5.0
Gel buffer (see *Table 1*)	3.0	4.0	6.0	3.75	5.0	5.8	6.8
Water	2.0	5.4	13.4	9.7	14.7	17.8	22.0
TEMED	0.025	0.025	0.025	0.025	0.025	0.025	0.025
10% (w/v) APS	0.25	0.25	0.25	0.25	0.25	0.25	0.25

Electrophoresis buffer: see *Table 1*.
[a]Columns represent volume (ml) of stock solutions required to prepare gel mixtures.
[b]For gels above 5% final acrylamide concentration, the bisacrylamide can be reduced to 0.375 g per 100 ml.

Table 3. Recipes for non-denaturing polyacrylamide slab gels for RNA separation

	Stacking gel[a,b]	Resolving gel[a]	
Stock solution	4%	4%	15%
Acrylamide–bisacrylamide (15:0.75)	2.5	10.7	—
Acrylamide–bisacrylamide (30:0.37)	—	—	20.0
Gel buffer (see *Table 1*)	0.38	8.0	8.0
Water	6.35	21.0	12.0
10% (w/v) APS	0.3	0.3	0.3
TEMED	0.03	0.03	0.03

Electrophoresis buffer: see *Table 1*.
[a]The columns represent volumes (ml) of stock solutions required to prepare gel mixtures.
[b]4% stacking gel for the top of the 15% slab resolving gel. The final buffer concn. in the stacking gel is one-fifth of that of the lower gel.

Table 4. Recipe for a 2.4–5.0% non-denaturing polyacrylamide gradient slab gel for RNA separation

Stock solution	% Acrylamide[a]	
	2.4	5.0
Acrylamide (30%)	2.0	4.16
Bisacrylamide (2%)	1.56	1.56
40% (w/v) sucrose	3.1	12.5
Gel buffer (see *Table 1*)	5.0	5.0
Water	13.3	1.8
10% (w/v) APS	0.0625	0.0625
10% (v/v) TEMED	0.0938	0.025

Electrophoresis buffer: see *Table 1*.
[a]The columns represent volumes (ml) of stock solutions required to prepare 25 ml gel mixture.

Table 5. Recipe for a 4–15% non-denaturing polyacrylamide gradient slab gel for RNA separation

Stock solution	% Acrylamide[a]	
	4.0	15.0
Acrylamide (30%)	2.67	10.0
Bisacrylamide (2%)	1.5	3.51
Buffer A[b]	3.0	—
60% (w/v) sucrose in buffer B[c]	1.67	6.67
Water	11.0	—
10% (w/v) APS	0.0625	0.0625
10% (v/v) TEMED	0.0938	0.025

Electrophoresis buffer: see *Table 1*.
[a]The columns represent volume (ml) of stock solutions required to prepare 20 ml of gel mixtures.
[b]Buffer A: gel buffer prepared as described in *Table 1*.
[c]Buffer B: 108 mM Tris base, 90 mM NaH_2PO_4, 3 mM EDTA.

Table 6. Denaturants used in denaturing gels for separating nucleic acids

Denaturant	Effective concn.	Comments
Alkali (>pH 12)	0.1 M	
Formamide	98%	For DNA only, RNA degraded; deaminates polyacrylamide
Methylmercuric acid	3–5 mM	Deionize before use; stops agarose gels setting
Urea	8 M at 60°C	Very toxic, use fume hood; run gels in borate sulfate buffer

Table 7. Recipes for denaturing formamide polyacrylamide gels for RNA separation

Constituent	% Acrylamide[a]		
	10.0	4.0	3.5
Acrylamide	2.4	0.91	0.75
Bisacrylamide	0.1	0.09	0.13
Diethylbarbituric acid[b]	0.092	0.092	0.092
TEMED[c]	0.06	0.06	0.06
Formamide (deionized)[c]	to make up to 25 ml	to make up to 25ml	to make up to 25 ml
18% (w/v) APS[c]	0.2	0.2	0.2

After mixing all the constituents, adjust the gel mixture to pH 9.0 with conc. HCl and adjust to the final 25 ml volume with the deionized formamide. The gels contain approximately 98% formamide. Where diethylbarbituric acid is used in the gels, the electrophoresis buffer is 20 mM NaCl in water. When diethylbarbituric acid is replaced with sodium phosphate, the buffer is 10 mM sodium phosphate in water adjusted to pH 6.0 or pH 9.0. 98% formamide has been used in the electrophoresis buffer reservoirs in place of water.

[a]Columns represent mass (g) of constituents required unless otherwise indicated.
[b]Can be replaced with 0.25 ml 1 M Na_2HPO_4, final pH adjusted to pH 9.0, or with 0.25 ml 1 M NaH_2PO_4 to give a final pH of pH 6.0. In each case, use 0.1 ml 36% (w/v) APS to compensate for the added water.
[c]Columns represent constituents required in volume (ml) to prepare the gel mixtures.

Table 8. Recipe for denaturing urea gels for RNA separation

Gel mixture:	As described in *Table 2*, with the exception that all the solutions used to make the gels (i.e. stock acrylamide, gel buffer and water) should be made up to contain a final concentration of 8 M urea.
Electrophoresis buffer:	Use any of the electrophoresis buffers described in *Table 1*; however, it is better to use a low ionic strength buffer to ensure denaturation of RNA. Prepare electrophoresis buffers with urea to a final concentration of 8 M.

Table 9. Recipe for denaturing methylmercuric hydroxide gels for RNA separation

Constituent	Concn.
Agarose	1.0%
Gel buffer	50 mM boric acid, 5 mM disodium tetraborate (decahydrate), 10 mM Na_2SO_4, 1 mM EDTA·Na_2·$2H_2O$, pH 8.19
Methylmercuric hydroxide	5 mM

Electrophoresis buffer: as described for gel buffer.

Table 10. Recipes for non-denaturing and denaturing agarose gels for RNA separation

Gel system	Gel mixtures
Non-denaturing systems	1.5% (w/v) agarose in buffers 2 and 4 given in *Table 1*.
Partially or completely denaturing buffer systems	1.5% (w/v) agarose with 50% formamide or with 6 M urea [1] or with 4 mM methylmercuric hydroxide [2].

Table 11. Recipe for horizontal slab agarose gel electrophoresis of RNA denatured with glyoxal and dimethylsulfoxide[a]

Gel mixture: 1% (w/v) agarose in 0.1 M sodium phosphate pH 7.0

Electrophoresis buffer: 10 mM sodium phosphate pH 7.0. As glyoxal dissociates from the RNA at pH greater than pH 8.0, it is necessary to recirculate the electrophoresis buffer constantly or change the buffer every 30 min during the run, to maintain the pH.

[a]For RNA up to 1 kb in size use 1.4% or 1.5% (w/v) agarose gels; if the RNA is larger try using 0.8–1.0% (w/v) agarose gels.

Table 12. Recipe for horizontal slab agarose gel electrophoresis of RNA denatured with formaldehyde and formamide[a]

Gel mixture: 1% (w/v) agarose in Mops–EDTA pH 7.0[b], 2.2 M formaldehyde

Electrophoresis buffer: Mops–EDTA pH 7.0. It is not necessary to recirculate the electrophoresis buffer constantly but if this is not done then it is advisable to change the buffer several times during the run.

[a] See footnote[a], *Table 11*.

[b] Mops–EDTA: 40 mM 3-(*N*-morpholino)-propanesulfonic acid (pH 7.0), 10 mM sodium acetate, 1 mM EDTA (pH 7.0).

Table 13. Buffers for non-denaturing gels for DNA separation[a]

Electrophoresis buffer	Comments
Tris–borate–EDTA (1 × TBE): 89 mM Tris–borate, 2.5 mM EDTA, pH 8.3 at 20°C	Popular use for polyacrylamide and some agarose gels. With some types of agarose, borate–agarose complexes may form and cause high electroendosmosis and gel damage. Tris base is oxidized to a UV-absorbing compound during electrophoresis.
50 mM Tris–acetate pH 7.5–8.0, or 50 mM sodium acetate–acetic acid pH 7.5–8.0	Acetate is oxidized to carbonate during electrophoresis, raising the pH; however, especially where DNA is to be recovered, acetate buffer is a good choice when combined with buffer recirculation or replacement. Tris base is oxidized to a UV-absorbing compound during electrophoresis.
50 mM Tris–NaH_2PO_4 pH 7.5–8.0, or 50 mM Na_2HPO_4–NaH_2PO_4 pH 7.5–8.0	The first has higher buffering capacity. Recirculation between the reservoirs may be required when using the second phosphate-based buffer. These buffers should be avoided when DNA is to be recovered by ethanol precipitation, as the phosphate will also precipitate from solutions at concentrations greater than 10 mM.

[a] 1–5 mM EDTA should be present in all buffers to chelate divalent cations.

Table 14. Recipes for non-denaturing polyacrylamide gels for DNA separation

Stock solution	% Acrylamide[a]		
	20	12	5
Acrylamide (40%)	50.0	30.0	12.5
Bisacrylamide (2%)	33.0	20.0	12.5
10 × TBE (see *Table 13*)	10.0	10.0	10.0
Water	5.95	38.95	63.95
10% (w/v) APS	1.0	1.0	1.0
TEMED	0.05	0.05	0.05

[a]The columns represent volume (ml) of stock solutions required to prepare 100 ml gel mixture.

Table 15. Buffers for denaturing gels for DNA separation[a]

Buffer system	Comments
30 mM NaOH, 2 mM EDTA	For agarose gels
5 mM methylmercuric hydroxide, 50 mM boric acid, 5 mM sodium borate, 25 mM Na_2SO_4, 1 mM EDTA pH 8.2	For agarose gels; recommended if absolute mol. wt determination in the absence of secondary structure is required
98% formamide containing 20 mM diethylbarbituric acid pH 9.0	For polyacrylamide gels
7 M urea, 2.5 mM EDTA, 89 mM Tris–borate pH 8.3	For polyacrylamide gels

[a]See also *Table 6*.

Table 16. Preparation of denaturing urea–polyacrylamide gels for DNA separation

Gel mixture: Prepare these as for the non-denaturing gels (*Table 14*). However, the volume of water is reduced to allow for the addition of 42 g urea to give a final 99 ml volume. Then 1 ml 10% (w/v) APS and 0.05 ml TEMED are added. The final gel mixture contains 7 M urea.

Gel Electrophoresis of Nucleic Acids

Table 17. Recipes for denaturing formamide polyacrylamide gels for DNA separation

Stock solution	% Acrylamide[a]	
	10	4
Acrylamide (40%)	25.0	10.0
Bisacrylamide (2%)	0.1	0.1
Diethylbarbituric acid[b]	0.37	0.37
Formamide (deionized)	80.0	80.0
10% (w/v) APS	1.5	1.5
TEMED	0.24	0.24

Adjust to pH 9.0 with 1 M HCl and add deionized formamide to a final 100 ml volume.

[a] The columns represent volume (ml) of constituents required to prepare 100 ml gel mixtures, unless otherwise indicated.

[b] Mass (g) of constituent required.

Table 19. Electrophoresis buffers for PFGE

Buffer	Preparation	Comments
0.5 × Tris–acetate–EDTA (TAE)	2.42 g Tris base, 0.571 ml glacial acetic acid and 2 ml 0.5 M EDTA, pH 8.0 per liter	Preferred over TBE
0.5 × Tris–borate–EDTA (TBE)	5.4 g Tris base, 2.75 g boric acid and 2 ml 0.5 M EDTA, pH 8.0 per liter	TBE buffer can lead to subsequent problems with transferring the DNA if blotting on to nitrocellulose filters

Table 18. Recipe for agarose gels for DNA separation

Gel mixture: 1% (w/v) agarose[a] in an appropriate buffer (see *Tables 13* and *15*)

[a] Alternatively prepare gel of desired agarose concentration.

Table 20. Recipes for electrophoresis of synthetic oligonucleotides

	% Acrylamide[a]		
Stock solution	20.0	12.0	8.0
Acrylamide–bisacrylamide (38:2)	75.0	45.0	30.0
10 × TBE (see *Table 13*)	15.0	15.0	15.0
Urea[b]	63.0	63.0	63.0
1.6% (w/v) APS	6.6	6.6	6.6
TEMED	0.1	0.1	0.1

Electrophoresis buffer: 1 × TBE buffer.
[a]The columns represent volume (ml) of constituents required to prepare gel mixtures, unless otherwise indicated.
[b]Mass (g) of constituent required.

Table 21. Recipe for the original procedure for oligonucleotide fractionation [3]

Stock solution	First dimension (10% acrylamide)[a]	Second dimension (20–23% acrylamide)[a]
Acrylamide–bisacrylamide (40:1.3)	37.5	75
9 M urea	100	—
1 M citric acid	3.75	—
1 M Tris pH 8.0[b]	—	6
2.5% (w/v) FeSO$_4$·7H$_2$O	0.6	—
10% (w/v) ascorbic acid	0.6	—
30% (w/v) H$_2$O$_2$	0.06	—
Water	7.49	68.35
10% (w/v) APS	—	0.6
TEMED	—	0.05

[a]The columns represent volume (ml) of stock solutions required to prepare 150 ml gel mixture.
[b]Adjusted to pH 8.0 with citric acid.

Table 22. Comparison of the original procedure and some modifications for oligonucleotide fractionation

Dimension	Gel concn. (% w/v)[a]		Buffer	Reference
	Acrylamide	Bisacrylamide		
First	10.0	0.325	0.025 M citric acid, 6 M urea	3
Second	20.0	0.65	0.04 M Tris–citrate pH 8.0	
First	10.0	0.35	0.025 M citric acid, 6 M urea	4
Second	20.0	0.66	0.1 M Tris–borate, 0.0025 M EDTA, pH 8.3	
First	10.0	0.325	0.025 M citric acid, 6 M urea	5
Second	20.0	0.65	0.2 M Tris–borate, 0.005 M EDTA, pH 8.3	
First	10.0	0.3	0.025 M citric acid, 6 M urea[b]	6
Second	22.8	0.8	0.05M Tris–borate, pH8.3	

[a]Amounts of catalyst used were essentially the same in all procedures (see *Table 21*).
[b]In this method, only the gel contained urea. The electrophoresis buffer used was 0.025 M citric acid.

Table 23. Electrophoresis of RNA fragments

| Dimension | Gel concn. (% w/v) | | Buffer | RNA fragments separated[a] | Reference |
	Acrylamide	Bisacrylamide			
First Second	12.1 12.1	0.4 0.4	0.04 M Tris–acetate pH 8.4 0.04 M Tris–acetate, 8 M urea, pH 8.4	$13 \leqslant N \leqslant 79$	7
First Second	10.0 20.0	0.325 0.65	0.025 M citric acid, 6 M urea 0.04 M Tris–citrate pH 8.0	$10 \leqslant N \leqslant 80$	3
First Second	8.0 16.0	0.26 0.52	0.025 M citric acid, 6 M urea 0.04 M Tris–citrate pH 8.0	$N > 80$	3

Equilibration solutions: water and 8 M urea. Sealing gel: gel solution containing no buffer.
In the first procedure, polymerization in both dimensions was achieved using 0.1 ml TEMED and 0.04% (w/v) APS. The gels for the second procedure were prepared as described in *Table 21*. Preparation of gels for the third procedure was exactly the same except that the acrylamide and bisacrylamide concentrations were a factor of 0.8 lower.
[a]N designates the chain length of separated RNA fragments.

Table 24. Electrophoresis of small RNA molecules

Dimension	Gel concn. (% w/v)		Buffer	Reference
	Acrylamide	Bisacrylamide		
First	9.5	0.5	0.089 M Tris base, 0.0025 M EDTA, 0.089 M H_3BO_3, pH 8.3	8[a]
Second	19.0	1.0	0.089 M Tris base, 0.0025 M EDTA, 0.089 M H_3BO_3, pH 8.3	
First	15.2	0.8	0.089 M Tris base, 0.0025 M EDTA, 0.089 M H_3BO_3, 7 M urea	9[b]
Second	15.2	0.8	0.089 M Tris base, 0.0025 M EDTA, 0.089 M H_3BO_3, pH 8.3	

[a]Polymerization for both dimensions was achieved using 0.4 ml diaminopropionitrile (DMAPN) and 0.04% (w/v) APS.
[b]Polymerization was achieved using 0.42 ml DMAPN and 0.056% (w/v) APS. In both dimensions a layer of approximately 1.5 cm of 'stacking gel' was cast on top of the resolving gel. It contained acrylamide–bisacrylamide (4.75:0.25) and 0.05 M Tris–HCl pH 3.6. To 5 ml of this polyacrylamide solution 0.25 ml 0.02% riboflavin was added.

Table 25. Electrophoresis of mRNA [10]

Dimension	Gel concn. (% w/v)		Buffer	Catalyst
	Acrylamide	Bisacrylamide		
First	6.0	0.16	0.04 M Tris base, 0.02 M sodium acetate, 0.001 M EDTA, acetic acid to pH 7.2	0.2% (w/v) APS, 0.2% (v/v) TEMED
Second	6.0	0.16	0.04 M Tris base, 0.02 M sodium acetate, 0.001 M EDTA, acetic acid to pH 7.2, 5 M urea	0.2% (w/v) APS, 0.2% (v/v) TEMED

Equilibration buffer: 0.02 M Tris base, 0.01 M sodium acetate, 0.5 mM EDTA, acetic acid to pH 7.2, 5 M urea.
Sealing gel: 1% (w/v) agarose in 0.02 M Tris base, 0.01 M sodium acetate, 0.5 mM EDTA, acetic acid to pH 7.2, 0.2% (w/v) SDS, 5 M urea.

Table 26. Recipe for two-dimensional gel electrophoresis of DNA restriction fragments

Gel mixture for both dimensions: 0.7% (w/v) agarose in 2 × TBE buffer (see *Table 13*)

Electrophoresis buffer: 2 × TBE buffer.
Sealing gel: 0.7% (w/v) agarose in 2 × TBE buffer.

Table 27. Recipes for separation of DNA fragments on denaturing gradient gels

Dimension	Gel mixture	Reference
First	1% (w/v) agarose gel in TAE buffer (40 mM Tris, 20 mM sodium acetate, 1 mM EDTA, pH 8.0)	11
Second	Low density: 4% acrylamide[a] in TAE buffer. Add 0.2 ml 20% (w/v) APS and 4 μl TEMED	
	High density: 4% acrylamide[a], 7 M urea, 40% (v/v) formamide in TAE buffer. Add 0.2 ml 20% (w/v) APS and 4 μl TEMED.	12

Electrophoresis buffer: 1 × TAE buffer.
Sealing gel: 1% (w/v) agarose gel in TAE buffer.
[a]4% acrylamide prepared from acrylamide–bisacrylamide (30:0.8) stock solution.

Table 28. Recipe for analysis of nucleic acid–protein interactions

Gel mixture: 25 ml 5% acrylamide solution in 1 × TBE buffer (see *Table 13*), 0.2 ml 10% (w/v) APS, 0.02 ml TEMED

Electrophoresis buffers: 1 × TBE buffer; 1 × TBE buffer containing 3 mM $MgCl_2$; 1 × TBE buffer containing 200 μM cAMP; 1 × TAE buffer (see *Table 27*); 10 mM Tris; or 2 M NaCl.
Sealing gel: 5 ml 5% stock acrylamide–bisacrylamide (30:1) in 1 × TBE buffer, 0.2 ml 10% (w/v) APS, 0.02 ml TEMED.

Table 29. Recipe for preparation of an agarose–acrylamide composite slab gel for analysis of DNA–protein complexes

Gel mixture: 15 ml 1% (w/v) agarose in 1 × TBE buffer (see *Table 13*), 5.2 ml 15% stock acrylamide–bisacrylamide solution (30:1) in 1 × TBE buffer, 10.0 ml water, 0.2 ml 10% (w/v) APS, 0.02 ml TEMED

Electrophoresis buffer: 1 × TBE buffer.

Table 30. Recipe for preparation of polyacrylamide gel for analysis of 5' and 3'-overhang end-labeled DNA

Gel mixture (5% acrylamide): 12.5 ml 30% stock acrylamide–bisacrylamide solution (29:1), 15 ml 5 × TBE buffer (see *Table 13*), 47.5 ml water, 0.4 ml 10% (w/v) APS, 0.03 TEMED.

Electrophoresis buffer: 1 × TBE buffer.

Table 31. Recipe for gel retardation analysis

Gel mixture (5% acrylamide): 12.5 ml 30% stock acrylamide–bisacrylamide solution (29:1), 3 ml 5 × TBE buffer (see *Table 13*), 59.5 ml water, 0.4 ml 10% (w/v) APS, 0.03 ml TEMED.

Electrophoresis buffer: 0.2 × TBE buffer.

Table 32. Recipe for polyacrylamide DNA sequencing gel electrophoresis

Gel mixture: 5 g acrylamide, 0.2 g bisacrylamide, 50 g urea, and 20 ml 5 × TBE buffer (see *Table 13*). Adjust to final 100 ml volume. Add 0.2 ml 10% (w/v) APS and 0.07 ml TEMED

Electrophoresis buffer: 1 × TBE buffer.

Table 33. Recipe for the preparation of buffer-gradient polyacrylamide gels for DNA sequencing

Stock solutions	6% Acrylamide/urea	
	Bottom solution	Top solution
6% acrylamide/urea bottom solution[a]	10.0	—
6% acrylamide/urea top solution[b]	—	35
10% (w/v) APS	0.04	0.12
TEMED	0.015	0.05

Electrophoresis buffer: 1 × TBE buffer (see *Table 13*).

[a]6% Acrylamide/urea bottom solution: 75ml 40% acrylamide–bisacrylamide (38:2), 50 ml 5 × TBE buffer, 230 g ultrapure urea. Adjust to 500 ml with deionized water.

[b]6% Acrylamide/urea top solution: 30 ml 40% acrylamide–bisacrylamide (38:2), 50 ml 5 × TBE buffer, 92 g ultrapure urea, 20 g sucrose, 10 mg bromophenol blue. Adjust to 500 ml with deionized water.

Table 34. Recipe for the preparation of a polyacrylamide slab gel for nucleosome separations [13]

Gel mixture (5% acrylamide): 12.5 ml 30% acrylamide–bisacrylamide stock solution (29:1), 7.5 ml 10 × TBE buffer (see *Table 13*), 54.7 ml water, 0.2 ml 10% (w/v) APS, 0.05 ml TEMED.

Electrophoresis buffer: 1 × TBE buffer.

Table 35. Recipe for the preparation of an agarose–polyacrylamide composite gel for nucleosome separation

Gel mixture (0.5% agarose–3.5% acrylamide): 8.6 ml 30% acrylamide–bisacrylamide stock solution (29:1), 10 ml 10 × TAE buffer (see *Table 27*), 30 g glycerol, adjusted to 50 ml with water, 0.1 ml 10% (w/v) APS, 0.025 ml TEMED. Mix polyacrylamide solution with 1% (w/v) agarose.

Electrophoresis buffer: 1 × TAE buffer.

Table 36. Recipe for second-dimensional polyacrylamide slab gel for the separation of native nucleosomal DNA

Gel mixture: 12.5 ml 30% acrylamide–bisacrylamide stock solution (29:1), 7.5 ml 10 × TBE buffer (see *Table 13*), 47.25 ml water, 7.5 ml 10% (w/v) SDS, 0.2 ml 10% (w/v) APS, 0.05 ml TEMED.

Equilibration buffer: 0.2 × TBE, 1% (w/v) SDS, 20% glycerol, 0.01% bromophenol blue.
Electrophoresis buffer: 1 × TBE buffer containing 1% (w/v) SDS.

Table 37. Recipe for second-dimensional polyacrylamide gel for the separation of denatured nucleosomal DNA

Gel mixture (6% acrylamide): 20 ml 30% acrylamide–bisacrylamide stock solution (29:1), 10 ml 10 × TBE buffer (see *Table 13*), 42 g urea, 37 ml water, 2 g SDS, 0.2 ml 10% (w/v) APS, 0.1 ml TEMED.

Equilibration solution: 9 M urea, 1% (w/v) SDS, 1 × TBE buffer.
Electrophoresis buffer: 1 × TBE buffer containing 2% (w/v) SDS.

Gel Electrophoresis of Nucleic Acids

Table 38. Recipes of gels used for the electrophoresis of polysomes and ribosomes

Gel composition (polyacrylamide/ agarose)	Gel buffer	Stock solutions									
		Agarose (g)	Water (ml)	20% Polyacrylamide solution[a] (ml)	6.4% DMAPN[b] (ml)	10× TBE[c] (ml)	1 M Tris–HCl (ml)	3 M KCl (ml)	1 M MgCl$_2$ (ml)	1.6% (w/v) APS (ml)	Refs
2.25%/0.5%	25 mM Tris–HCl 60 mM KCl 10 mM MgCl$_2$ pH 7.6	0.8	118	18	10	—	4	3.2	1.6	5	14
2.25%/0.5%	25 mM Tris–HCl 6 mM KCl 2 mM MgCl$_2$ pH 7.6	0.8	122	18	10	—	4	0.32	0.32	5	15
2.75%/0.5%	25 mM Tris–HCl 0.2 mM MgCl$_2$ pH 7.6	0.8	119	22	10	—	4	—	0.032	5	16
3%/0.5%	pH 8.3	0.8	105	24	10	16	—	—	—	5	17, 18

Electrophoresis buffer: same as gel buffer.

[a] 20% polyacrylamide solution: 19% acrylamide, 1% bisacrylamide.

[b] DMAPN, 3-dimethylaminopropionitrile.

[c] 10 × TBE: prepared at ten times the concentration described in *Table 13*.

Table 39. Non-radioactive labels and labeling methods

Label	Labeling methods[a]	Label density	Detection system[b]	Detection method
Biotinylated NTP or dNTP	A B	5% 0.1–1%	AP/NBT AP/NBT HRP/DAB HRP/ECL Fluorescein	Color Light Color Light Fluorescence
Digoxygenin-conjugated NTP or dNTP	A B	5% 0.1–1%	AP/NTB AP/DOP HRP/DAB HRP/ECL Fluorescein	Color Light Color Light Fluorescence
HRP–PEI conjugate[b]	C	1–5%	DAB ECL	Color Light
Photobiotin	C	5%	AP/NBT AP/DOP HRP/DAB HRP/ECL Fluorescein	Color Light Color Light Fluorescence

[a]Labeling methods: A; nick translation, random-primer labeling, single-primer labeling, phage polymerase transcription, direct biotinylation, and direct enzyme labeling. B; single-primer labeling, phage polymerase transcription, end-filling/substitution, kinasing, tailing, and ligation. C; direct conjugation.

[b]Abbreviations: AP, alkaline phosphatase; DAB, diaminobenzidine; DOP, dioxetane phosphate; ECL, enhanced chemiluminescence; HRP, horse-radish peroxidase; NBT, nitroblue tetrazolium; PEI, polyethyleneimine.

Table 40. Radioactive labeling methods

Radionuclide	Specific activity of labeled nucleotides (TBq)	Labeling methods	Specific activity of probe (d.p.m. μg^{-1}) ($\times 10^8$)
^{3}H	0.9–3.7	Nick-translation	0.5
		Random-priming	1.5
		Phage RNA polymerase	0.5
^{32}P	14.8–222.2	Nick-translation	5
		Random-priming	50
		Phage RNA polymerase	13
		End-labeling	0.05
^{35}S	14.8–55.5	Nick-translation	1
		Random-priming	7
		Phage RNA polymerase	13
^{125}I	37.0–74.0	Nick-translation	1
		Random-priming	15
		Phage RNA polymerase	10
		Direct iodination	2

Table 41. Sources of radioactive labels

Label	Source
Nucleosides	AIP, ICN, NEN, SCC
Nucleotides	AIP, ICN, NEN

Table 42. Inhibitors of nucleases

Inhibitor	Mol. wt	Solubility	Additional information
Actin	—	Sol. in water	Inhibits DNase I
Actinomycin D	1255.5	Sol. in water, EtOH and glycols	Inhibits DNase I and also RNA formation *in vivo*
Adriamycin	543.5	Sol. in water, MeOH and alcohols	Inhibits DNase I
Aurintricarboxylic acid	473.4	Sol. in water	Inhibits DNase I, RNase I, S1 nuclease and exonuclease III
Daunomycin	527.5	Sol. in water and MeOH	Inhibits DNase I
Diethyl pyrocarbonate	162.1	Immiscible in water	Alkylating agent that non-specifically modifies any proteins in the sample
Iodoacetamide	233.0	Sol. in water	As described for diethyl pyrocarbonate
Iodoacetic acid	186.0	Sol. in water	As described for diethyl pyrocarbonate
Nucleoside phosphonic acids	—	—	Inhibition of micrococcal nuclease
RNase inhibitor	—	Sol. in water	Inhibits ribonucleases
Vanadyl ribonucleosides	—	—	Inhibition of micrococcal nuclease, polynucleotide phosphorylase, restriction enzymes and translation ribonucleases A, N1, T1, U2

Table 43. Tracking dyes for nucleic acids

Tracking dye	Application
Bromocresol green	DNA agarose electrophoresis
Methylene blue	RNA, RNase stain
Methyl green	Native DNA, acidic or neutral tracking dye
Pyronine Y	RNA, acidic tracking dye
Toluidine blue O	RNA, RNase stain
Xylene cyanol FF	DNA sequencing

Table 44. Sample loading buffers for gel electrophoresis of nucleic acids

Application	Sample loading buffer	Cross-references[a]
Non-denaturing gels for RNA separation	36 mM Tris base, 30 mM NaH$_2$PO$_4$, 1 mM EDTA, 0.2% (w/v) SDS, pH 7.7, 5–15% (w/v) sucrose	2–5, 10
Denaturing formamide gels for RNA separation	100% Buffered formamide containing 10% (w/v) sucrose	7, 10
Denaturing urea gels for RNA separation	8 M Urea, 10% (w/v) sucrose	8, 10
Denaturing methylmercuric hydroxide gels for RNA separation	0.0125 M Methylmercuric hydroxide, 50 mM boric acid, 5 mM disodium tetraborate (decahydrate), 10 mM Na$_2$SO$_4$, 1 mM EDTA·Na$_2$·2H$_2$O, pH 8.19, 0.1% bromophenol blue.	9, 10
Horizontal slab agarose gel electrophoresis of RNA denatured with glyoxal and dimethylsulfoxide	50% Glycerol, 0.01 M sodium phosphate pH 7.0, 0.4% bromophenol blue	11
Horizontal slab agarose gel electrophoresis of RNA denatured with formaldehyde and formamide	50% Glycerol, 1 mM EDTA, 0.4% bromophenol blue	12
Electrophoresis of DNA	15% (w/v) Ficoll, 0.25% Orange G, 250 mM EDTA[b]	14
Denaturing formamide polyacrylamide gels for DNA separation	Samples dissolved in 0.023 (w/v) diethylbarbituric acid in deionized formamide (adjusted to pH 9.0 with 1 M NaOH); add one-tenth volume of 20% (w/v) Ficoll, 0.25% bromophenol blue	17
Electrophoresis of synthetic oligonucleotides	98% Formamide–water[c] or 7 M urea[d]	20
Two-dimensional separation of RNA	Purified RNA is dissolved in 8 µl 6 M urea; add 2 µl of mixture containing 300 mg urea, 500 mg sucrose, 2 mg bromophenol blue, 2 mg xylene cyanol FF and 5 mg trypan red per milliliter[e]	23–25
Two-dimensional separation of DNA	25% Glycerol, 60 mM EDTA, 0.1% bromophenol blue	26, 27

Continued

Table 44. Sample loading buffers for gel electrophoresis of nucleic acids, *continued*

Application	Sample loading buffer	Cross-references[a]
Analysis of nucleic acid–protein interactions	2.5% (w/v) Ficoll, 0.005% xylene cyanol FF, 0.005% bromophenol blue	28
Analysis of 5′ and 3′-overhang end-labeled DNA	23 mM EDTA pH 8.0, 2.3% (w/v) SDS, 11% (w/v) Ficoll, 0.05% bromophenol blue	30
Gel retardation analysis	20 mM Hepes–NaOH pH 7.6, 4% (w/v) Ficoll, 5 mM MgCl₂, 40 mM NaCl, 0.1 mM EDTA, 0.5 mM DTT	31
DNA sequencing	95% (v/v) Deionized formamide, 0.02% xylene cyanol FF, 0.025% bromophenol blue	32, 33
Electrophoresis of nucleosomes using polyacrylamide slab gels	20% Glycerol or 4% (w/v) Ficoll, 0.01% bromophenol blue in 1 × TBE buffer (see *Table 13*).	34
Electrophoresis of nucleosomes using agarose–polyacrylamide gels	10% Glycerol, 2 mM Tris–HCl, 2 mM EDTA, pH 8.0	35
Electrophoresis of polysomes and ribosomes	0.5% (w/v) Agarose in 60 mM KCl, 10 mM MgCl₂, in 25 mM Tris–HCl pH 7.6.	38

[a]The numbers listed in this column indicate the tables to which the reader should refer for further details regarding the gel system used in conjunction with the sample loading buffers.

[b]Orange G can be replaced with 0.025% bromophenol blue, 0.025% bromocresol green and 0.025% xylene cyanol FF. Ficoll can be replaced with 30% (w/v) glycerol or 40% (w/v) sucrose.

[c]For small oligodeoxynucleotides (< 30 nucleotides).

[d]For longer oligodeoxynucleotides and short oligodeoxynucleotides with a high G or G + C content.

[e]If the first-dimensional gel contains no urea, then the RNA may be dissolved in water and urea may be omitted from the sample buffer.

Table 45. RNA size markers

RNA species	Source	Mol. wt ($\times 10^6$)[a]	No. of nucleotides
4S RNA	*Aspergillus*	0.0263	85
5S RNA	*E. coli*	0.0372	120
5.8S RNA	*Aspergillus*	0.0489	158
Histone H4 mRNA	Sea urchin	0.13	410
α-Globin mRNA	Rabbit	0.20	630
α-Globin mRNA	Mouse	0.22	696
α-Globin mRNA	Rabbit	0.22	710
α-Globin mRNA	Mouse	0.24	783
Immunoglobulin light chain mRNA	Mouse	0.39	1250
A2 crystallin mRNA	Calf lens	0.45	1460
16S rRNA	*E. coli*	0.53	1776
17S rRNA	*Aspergillus*	0.62	2000
18S rRNA	HeLa	0.71	2366
23S rRNA	*E. coli*	1.07	3566
25S rRNA	*Aspergillus*	1.24	4000
28S rRNA	HeLa	1.90	6333
Myosin heavy chain mRNA	Chicken	2.02	6500
Fibroin mRNA	Silkworm	57.0	19×10^3

[a]Molecular weights are approximate only and based upon an average 'molecular weight' of 310 for each nucleotide.

Table 46. Sizes of the restriction fragments of pAT153[a]

HincII	NarI	HgiDI	TaqI	HaeII	HinfI	Sau96	Sau3A	MnlI	HaeIII	HpaII	HhaI
2551	2866	1994	1444	1876	1631	1224	876	591	587	622	393
1106	657	657	602	439	517	616	665	400	458	492	337
	113	490	475	430	396	352	358	247	434	404	332
	21	382	368	370	298	274	341	218	339	242	270
		113	315	227	221	249	317	206	267	238	259
		21	312	181	220	222	272	206	234	217	206
			141	60	154	191	258	206	213	201	190
				53	145	179	105	204	192	190	174
				21	75	124	91	200	184	160	153
						88	78	179	124	160	152
						79	75	166	123	147	151
						42	46	156	104	122	132
						17	36	150	89	110	131
							31	96	80	90	109
							27	88	64	76	100
							18	81	57	67	93
							17	77	51	34	75
							15	61	21	26	67
							12	60	18	26	62
							11	38	11	15	60
							8	27	7	9	53
										9	40
											36
											33
											28
											21

Table 47. Sizes of the restriction fragments of phage λ cI *ts* 857[a]

XbaI	SalI	AvrII	SmaI	PvuI	EcoRI	BamHI	BglII	HindIII	AvaI	HpaI	PvuII
24508	32745	24322	19399	14321	21226	16841	22010	23130	14677	8666	21088
23994	15258	24106	12220	12712	7421	7233	13286	9416	8614	6911	4421
	499	74	8612	11936	5804	6770	9688	6557	6888	5414	4268
			8271	9533	5643	6527	2322	4361	4720	4535	4194
					4878	5626	2027	2322	4716	4491	3916
					3530	5505	564	2027	3730	4347	3638
							125	564	1881	3408	2296
								125	1674	3384	1708
									1602	3042	636
										2240	579
										734	532
										441	468
										410	343
										251	211
										228	141
											64

[a]See footnote[a], *Table 46.*

Gel Electrophoresis of Nucleic Acids

Table 48. Sizes of the restriction fragments of phage M13 mp7[a]

EcoRI	PvuII	XhoI	HaeII	EcoRII	KpnI	TaqI	HaeIII	RsaI	HpaII	AluI	EcoRI*
7196	6835	4020	3514	3975	1694	1018	2527	1345	1596	1446	1402
42	310	2535	2520	1809	1241	971	1623	1334	829	1330	665
		659	434	952	1226	927	1004	1004	818	600	567
		24	433	179	792	703	341	742	651	555	508
			329	139	666	639	311	604	545	484	416
			8	127	332	612	309	522	543	336	406
				57	318	579	245	322	472	331	323
					301	564	214	258	454	313	283
					196	441	169	201	357	220	272
					196	381	158	190	183	204	247
					163	239	117	163	176	201	213
					113	152	106	143	156	180	209
						12	102	107	130	159	176
							98	102	123	151	174
							69	93	79	140	152
								65	60	111	142
								27	30	111	109
								16	18	104	109
										93	102
										72	99
										63	96
										39	88
										27	76

26	69
24	63
	55
	53
	41
	40
	36
	33
	12
	1
	1

[a]See footnote[a], *Table 46.*

*Denotes 'star' activity. It has been demonstrated that under extreme non-standard conditions restriction endonucleases are capable of cleaving sequences which are similar but not identical to their defined recognition sequence. This altered specificity is termed 'star' activity. Under conditions of elevated pH and low ionic strength, *Eco*RI cleaves the sequence N/AATTN, while *Eco*RI* (*Eco*RI star activity) cleaves any site which differs from the accepted recognition sequence by a single base substitution, providing the substitution does not result in an 'A' to 'T' or a 'T' to 'A' change in the central (AATT) tetranucleotide sequence.

Table 49. Commercially available nucleic acid size markers[a]

Type of marker	Source
RNA markers	BDH, BHM, GBL, HSI, IBI, PMB, PML, SCC
DNA markers	APP, BHM, BRL, GBL, HSI, IBI, NBL, PMB, PML, SCC, STG
Pulse field markers	BHM, BRL, FIL, PMB, PML, SCC, STG
Oligonucleotide sizing markers	PMB

[a]Some of the suppliers offer prestained or radiolabeled size markers.

Gel Electrophoresis

Table 50. Staining procedures for detecting nucleic acids in gels

Staining technique	Destaining technique	Comments
Stain overnight in 0.2% methylene blue in 0.2 M sodium acetate buffer pH 4.7	Destain by frequent washes with water or in continuous running water	Methylene blue can be replaced by toluidine blue O, thionin and azure A
Soak overnight in 0.1% pyronine Y in 0.5% (v/v) acetic acid, 1 mM citric acid	0.5% Acetic acid	Use 10% acetic acid for destain when staining for RNA separated using formamide gels
Stain for 30–60 min with 1–5 µg ml^{-1} ethidium bromide in 0.5 M ammonium acetate	Wash the gel in distilled water or 0.5 M ammonium acetate for up to 2 h in the dark	RNA and DNA fluoresce pink; prolonged exposure to UV light can damage the nucleic acids
Immerse gel in 0.5 µg ml^{-1} ethidium bromide in water for 30 min	Destain in 1 × TBE buffer (see *Table 13*) for 30 min	
Fix in 10% acetic acid for 20 min, wash in 10% EtOH for 5 min, and pretreat gel in 0.1% (w/v) $K_2Cr_2O_7$, 0.02% (v/v) HNO_3 for 5 min; rinse gel in water for 2 min, three times; soak in 0.1% (w/v) $AgNO_3$, 0.06% (v/v) HCHO for 30 min	Rinse in water for 20 sec; develop in 3% (w/v) Na_2CO_3, 0.06% (v/v) HCHO, 0.0002% (w/v) $Na_2SO_3 \cdot 5H_2O$ for 2–5 min. Stop reaction by soaking gel in 10% acetic acid for 5 min	

Gels containing high molecular weight RNA or DNA can be stained directly, while gels with lower molecular weight nucleic acids should be fixed in 1 M acetic acid or 10% trichloroacetic acid for 30–60 min prior to staining.

Table 52. Transfer buffer solutions for nucleic acids

Transfer buffer	Application
20 × SSC[a]	Blotting of RNA and DNA from gels to nitrocellulose or nylon membranes
10 × SSC[a]	Transfer of PFGE-DNA to nylon membrane
Alkaline transfer buffer: 0.4 M NaOH	Transfer of DNA and PFGE-DNA to nylon membrane
2 × TBE[b]	DNA transfer to DEAE-cellulose membranes
1 × TBE[b]	Hybridization analysis of nucleosomal DNA separated on a two-dimensional denaturing polyacrylamide gel

[a]1 × SSC: 0.15 M NaCl, 0.015 M trisodium citrate, pH 7.0. Prepare to desired concentration.
[b]TBE buffer: as described in *Table 13*. Prepare to desired concentration.

Table 51. Denaturation of DNA in gels for transfer to nitrocellulose membranes

Solution for denaturation of DNA:	1.5 M NaCl, 0.5 M NaOH
Gel neutralizing solution:	3 M NaCl, 1 M Tris–HCl pH 8.0

Gel Electrophoresis of Nucleic Acids

Table 53. Prehybridization and hybridization solutions

Solution	Application
50% Formamide, $1 \times$ Denhardt's solution[a], $5 \times$ SSC[b], 1–2 mg sonicated denatured heterologous DNA ml^{-1} and 1% glycine	Prehybridization solution for hybridization of radioactive DNA to RNA attached to DBM-paper
50% Formamide, $1 \times$ Denhardt's solution[a], $5 \times$ SSC[b], 1–2 mg sonicated denatured heterologous DNA ml^{-1}	Hybridization solution for hybridization of radioactive DNA to RNA attached to DBM-paper
$5 \times$ Denhardt's solution[a] in $5 \times$ SSC[b], 10–50 µg ml^{-1} heterologous DNA	Prehybridization solution for filter hybridization of DNA
$5 \times$ Denhardt's solution[a] in $5 \times$ SSC[b], 0.1% (w/v) SDS, 10% dextran sulfate; 7% (w/v) SDS, 1 mM EDTA, 0.5 M Na$_2$HPO$_4$, pH 7.2; or BLOTTO[d]	Hybridization solutions for filter hybridization of DNA
$5 \times$ SET[c], $5 \times$ Denhardt's solution[a], 0.1% (w/v) SDS, 0.1% sodium pyrophosphate	Prehybridization solution for hybridization to nitrocellulose membranes after PFGE
$5 \times$ SET[c], $5 \times$ Denhardt's solution[a], 0.1% (w/v) SDS, 0.1% sodium pyrophosphate, 10% dextran sulfate	Hybridization solution for hybridization to nitrocellulose membranes after PFGE
$5 \times$ Denhardt's solution[a], 0.5 M phosphate buffer pH 7.5, 1% (w/v) SDS, 10% dextran sulfate, 100 µg ml^{-1} denatured sonicated salmon sperm DNA (optional)	Prehybridization and hybridization solution for hybridization to nylon membrane filters after PFGE

[a]Denhardt's solution: 0.02% (w/v) Ficoll, 0.02% polyvinylpyrrolidone, 0.02% (w/v) BSA. Prepare to desired concentration.
[b]SSC buffer: as described in *Table 52*. Prepare five times the concentration described.
[c]$5 \times$ SET: 0.75 M NaCl, 5 mM EDTA, 0.1 M Tris–HCl, pH 7.8.
[d]Stocked as a solution of 10% (w/v) non-fat powdered milk containing 0.2% sodium azide, stored at 4°C.

Table 54. Choice of isotope detection method [19]

Isotope	Matrix	Detection method
^{3}H	Agarose gels, polyacrylamide gels, or any membrane or paper filters	Fluorography
^{14}C/^{35}S	Agarose or polyacrylamide gels	Fluorography
	Any membrane or paper filters	Direct autoradiography
^{32}P/^{125}I	Agarose gels, polyacrylamide gels, or any membrane or paper filters	Autoradiography with intensifying screen or direct autoradiography

Table 55. Preparation of gels for fluorography

Preparation	Modifications
Soak polyacrylamide gel in approximately 20 times its volume of DMSO for 30 min. Soak gel in four volumes of 20% (w/v) PPO dissolved in DMSO for 3 h. Wash gel in 20 volumes of water for at least 1 h to remove DMSO and precipitate the PPO in the gel[a]. Dry gel under vacuum [20, 21].	100% Acetic acid used in place of DMSO as PPO solvent, since this can be used for both agarose and polyacrylamide gels [22]. 100% Methanol can also be used in place of DMSO for agarose gels, or ⩽2% agarose–acrylamide composite gels which do not change volume in methanol [21].
Fix gel in 5% trichloroacetic acid or methanol:acetic acid:water (5:1:5). Soak gel in 20 volumes of water for 30 min to remove the acid[b]. Soak gel in 10 volumes of 1 M sodium salicylate pH 7.0 for 20 min. Dry under vacuum [23].	5% Methanol and 1% glycerol have been added to the sodium salicylate to facilitate subsequent drying of the gel.

[a] A longer washing period may be necessary if artifactual blackening of the film by residual DMSO occurs.
[b] This step is necessary to prevent the sodium salicylate precipitating.

Table 56. Solutions used to extract and elute nucleic acids from agarose and polyacrylamide gels

Extraction/elution buffers	Application
Water-saturated butanol containing hexadecyl-trimethyl ammonium bromide (HTAB)	Extraction of RNA from agarose gels
0.2 M NaCl	Removal of RNA from butanol extracts
'880' Ammonia solution or 10% piperidine containing 1 mM EDTA; add water afterwards to swell gel slices and dissolve the hydrolyzed RNA	Elution of radioactively labeled RNA from polyacrylamide gel slices
$2 \times$ SSC and $6 \times$ SSC[a] with or without 0.2% (w/v) SDS; 0.15 M sodium acetate adjusted to pH 6.0 with acetic acid and containing 0.5% (w/v) SDS; or 0.6 M or 0.8 M lithium acetate pH 6.0 containing 0.5% (w/v) SDS	Elution of RNA from polyacrylamide gel slices
0.5 M Tris–HCl pH 7.0, 0.1% (w/v) SDS, 0.1 mM Na_2EDTA, 1 mM $MgCl_2$	Elution of RNA from polyacrylamide gels by diffusion; use for RNAs up to approximately 300 bases
0.001 M EDTA, 3 M sodium acetate, 0.1 M magnesium acetate, 0.001 M EDTA pH 6.0, carrier RNA	Extraction of RNA from two-dimensional gel electrophoresis; suitable for extraction from thin gel discs containing oligonucleotides, which diffuse relatively fast
1 M NaCl, carrier RNA	Extraction of RNA from two-dimensional gel electrophoresis; suitable for larger RNA fragments and thicker gel pieces
25 mM Sodium acetate buffer pH 8.0	Elution of RNA separated by two-dimensional gel electrophoresis Elution of DNA on to dialysis tubing
TAE buffer[b] containing 0.1 M NaCl and phenol	Extraction of DNA from agarose gels

200 mM NaCl and redistilled phenol	Isolation of DNA from low gelling temperature agarose
Low salt buffer: 0.2 M NaCl, 20 mM Tris–HCl pH 7.6, 1 mM EDTA	Extraction of DNA from polyacrylamide gels
High salt buffer: 1 M LiCl, 20 mM Tris–HCl pH 7.6, 1 mM EDTA	Elution of DNA, extracted from polyacrylamide gels, from isolation column
0.5 M NaCl; 2 M triethylammonium acetate; 50 mM triethylammonium acetate; water; 0.5 M NaCl, 0.1 M Tris–HCl pH 7.0 containing 1 mM EDTA	Extraction of synthetic oligonucleotides from gels
Low salt buffer: 0.5 × TBE buffer[c]	Elution of synthetic oligonucleotides from gels
High salt buffer: 10 M ammonium acetate, 0.01% bromophenol blue	

[a] SSC buffer: see *Table 52*. Prepare to desired concentration.
[b] TAE buffer: see *Table 27*.
[c] TBE buffer: see *Table 13*. Prepare to desired concentration.

Gel Electrophoresis of Nucleic Acids

Chapter 6 **GEL ELECTROPHORESIS OF CARBOHYDRATES**

Many complex carbohydrates which exist as glycoconjugates are known to have important biological functions. There has been considerable research into their chemical structures. Structural analysis has been performed by using classical chemical procedures such as mass spectroscopy and nuclear magnetic resonance. However, they have significant limitations. Consequently complementary methods have been developed. Enzymological degradation of oligosaccharides by specific glycosidases has allowed analysis of smaller quantities of saccharides. This technique has been enhanced by the derivatization of saccharides with fluorophores and the ability to separate the products with high resolution by PAGE (*Table 1*). *Table 2* compares fluorophores used for labeling oligosaccharides.

Table 1. Recipes for oligosaccharide gel electrophoresis [1, 2]

Gel mixture	Electrophoresis buffer
20–40% (w/v) acrylamide containing 0.67–1.06% (w/v) bisacrylamide, respectively, 0.016% (w/v) APS, 0.08% (v/v) TEMED	0.025 M Tris, 0.192 M glycine, pH 8.3
20% (w/v) acrylamide containing 0.67% (w/v) bisacrylamide, 0.1% (w/v) APS, 0.1% (v/v) TEMED	Either 0.1 M Tris–borate pH 8.3 or 0.025 M Tris, 0.192 M glycine, pH 8.3

Table 2. Fluorophores used to label saccharides for electrophoretic separation [1, 2]

8-Aminonaphthalene-1,3,6-trisulfonic acid (ANTS)	2-Aminoacridone
Dissolved in glacial acetic acid/water (3:17)	Dissolved in glacial acetic acid/DMSO (3:17)
ANTS derivatization imparts both charge and fluorescence to the saccharides; labeled saccharides have overall negative charge	2-Aminoacridone confers no charge on the labeled saccharides at the pH of the electrophoretic system gel
As little as 1 pmol of ANTS-labeled saccharide detected photographically when gels illuminated by UV light	As little as 0.63 pmol of 2-aminoacridone-labeled saccharide detected photographically when gels illuminated by UV light
As little as 0.2 pmol of ANTS-labeled saccharide detected using a cool charge-coupled device	As little as 0.2 pmol of 2-aminoacridone-labeled saccharide detected using a cool charge-coupled device
Electrophoretic mobilities related largely to M_r of compounds but also influenced by individual chemical structures of the saccharides	Mobilities dependent partly on the size of each saccharide molecule but influenced strongly by various molecular structures
Neutral and acidic saccharide derivatives all negatively charged by the ANTS molecule and in turn they are not readily distinguishable electrophoretically as two groups of molecules	As no charge conferred by 2-aminoacridone it is possible to distinguish neutral and acidic saccharide derivatives

Gel Electrophoresis of Carbohydrates

Chapter 7 **TROUBLESHOOTING**

Table 1 lists symptoms, causes and appropriate remedies suggested for problems experienced in electrophoretic analysis of proteins. Symptoms 1–16 apply to one-dimensional gel electrophoresis, 17–20 to IEF, 21–26 to IPG and 27–28 to two-dimensional gel electrophoresis. However, many of the symptoms and remedies may apply to more than one electrophoresis system. Symptoms 29–39 are problems experienced with protein detection methods.

Table 2 lists symptoms, causes and appropriate remedies suggested for problems experienced in electrophoretic analysis of nucleic acids. Symptoms 1–11 apply to gel electrophoresis of RNA, 12–20 to DNA, 21–22 to oligonucleotides, 23 to gel retardation analysis and 24–26 to polysomes and ribosomes. Many of the symptoms and remedies may apply to more than one electrophoresis system. Symptoms 27–31 are problems experienced with autoradiography.

Table 1. Troubleshooting guide for proteins

Symptom	Cause	Remedy
1. Leaking mold	Dust or gel fragments on the gasket	Carefully clean gel plate and gasket
2. Lack of polymerization. The gel consistency is not firm, gel does not hold its shape after removal from the mold	Incorrect concentrations of prepared reagents; omission of reagent from gel mixture; impurities in reagents; old APS stock solution	Discard solutions and prepare fresh batch using pure reagents. Check that recommended polymerization conditions are used
3. Too fast or slow polymerization		Vary concentrations of polymerization catalysts

4. Cracking of gel during polymerization (usually only high concn. gels)	Excessive heat production by polymerization reaction	Use cooled solutions. For rod gels, siliconizing the tubes may also help
5. Gel cracking during electrophoresis.	Excessive current input overheating the gel	Use less current over a longer period of time
6. Detachment of slab gels from glass plates during gel electrophoresis	Inadequately cleaned plates; low concn. gels sometimes detach from rod gel tubes even though these are clean	Thoroughly clean plates and tubes; a solution of a strong decontaminating detergent such as Contrad 70, DECON 90, or RBS 35 should be used. Rinse extensively with distilled water, and finally with ethanol prior to air-drying. Alternatively, attach a piece of nylon mesh to bottom of the tube
7. Failure of the sample to form a layer at the well bottom when applied to slab gels	Omission of sucrose or glycerol from sample buffer; use of sample comb where teeth do not form a snug fit with the glass plates, permitting gel to polymerize between the teeth and glass plates which in turn interferes with sample loading	Prepare fresh batch of sample buffer. Use better fitting comb, but in the short term, excess gel removed from wells using syringe needle
8. Insoluble material in the sample	Provided that the ionic strength of non-dissociating buffers is high enough to prevent aggregation of native proteins, denatured protein represents the insoluble material	Removed by centrifugation prior to electrophoresis
9. Insolubility in SDS-containing buffers	Too little SDS; too little reducing agent; too low a pH, especially after TCA precipitation of proteins	Adjust concentrations of the said reagents; add urea in addition to SDS to ensure solubilization

Continued

Table 1. Troubleshooting guide for proteins, *continued*

Symptom	Cause	Remedy
10. Protein streaking along individual rod gels or slab gel tracks and protein at the gel origin	Protein precipitation followed by dissolution of the protein precipitates during electrophoresis; overloading of the gel	See symptoms no. 8 and no. 15; amount of sample loaded should be decreased
11. Protein bands observed in all tracks of a slab gel or all rod gels; continuous stained region from the gel origin to near the buffer front, even in tracks which have not been loaded with sample; same protein bands observed in several neighboring lanes of a slab gel	Contamination of the sample buffer; contaminated electrophoresis buffer; sample from one well has contaminated adjacent wells, usually by overflowing	Prepare fresh buffers; reduce the volume of sample loaded
12. High background of protein staining along individual rod gels or slab gel tracks with indistinct bands	Extensive sample proteolysis; use of impure grades of SDS	Working at low temperature and use of protease inhibitors during sample preparation. If problem occurs with SDS–PAGE, check that sample is heated to at least 90°C for 2 min during dissociation; use purified SDS
13. Distorted bands	Insoluble material; bubbles in the gel; inconsistent pore size throughout the gel; uneven heating of the gel	Filter gel reagents before use and ensure gel mixture is well mixed and degassed before pouring the gel; use a cooled apparatus or reduce the current at which electrophoresis is performed

the width of a stained band	sample accumulating at the low points prior to electrophoresis	gels in vibration-free places
15. Heavily stained band at the gel origin	Inability of a substantial portion of the protein to enter the resolving gel; may be aggregated protein in sample prior to electrophoresis or, in the case of non-denaturing discontinuous buffer systems, precipitation of the proteins due to the formation of highly concentrated zones during electrophoresis in the stacking gel	Aggregated protein (see symptom no. 8). In the case of the non-dissociating discontinuous buffer systems, use less concentrated samples with a continuous buffer system
16. Irreproducibility of protein band pattern; reduction in the staining intensity or complete loss of individual components or appearance of previously unobserved fast migrating bands	Caused by problems of sample preparation; proteolysis	See symptom no. 12
17. Waviness of bands near the anode	Carbonization of the catholyte; excess catalysts; too long sample slots; too low a concentration of carrier ampholytes	Prepare fresh NaOH with degassed distilled water and store properly in sealed plastic container; reduce amount of APS; fill slots with dilute carrier ampholytes; check the gel formulation To alleviate the problem, add low concentrations of sucrose, glycine or urea, and apply sample near the cathode. To salvage gel during the run,

Continued

Table 1. Troubleshooting guide for proteins, *continued*

Symptom	Cause	Remedy
		as soon as waves appear apply a new anodic strip soaked with a weaker acid (such as acetic acid instead of phosphoric acid) inside the original one, and move electrodes closer to one another
18. Burning along the cathodic strip	The formation of a zone of pure water at pH 7.0; hydrolysis of the acrylamide matrix after prolonged exposure to alkaline pH	Add to the acidic pH range a 10% solution of either the 3–10 or the 6–8 range ampholytes; choose a weaker base, if adequate, and, unless a pre-run of the gel is strictly required, apply electrode strips after loading samples
19. pH gradients different from expected	Cathodic drift; a large amount of a weak acid or base, supplied as sample buffer, may shift the pH range (e.g. 2-mercaptoethanol); addition of urea increases apparent pIs of the carrier ampholytes	Reduce running time to the required minimum (as experimentally determined for the protein of interest, or for a colored marker of similar molecular mass); increase viscosity of the medium with sucrose, glycerol, or urea; reduce the amount of APS; remove acrylic acid impurities by recrystallizing acrylamide and bisacrylamide; incorporate into the gel matrix a reactive base, e.g. 2-dimethylamino-propyl-methacrylamide

20. Sample precipitation at the application point	Protein aggregation	[1]. For acidic and alkaline pH ranges, problem alleviated by choice of anolytes and catholytes whose pH is close to the extremes of the pH gradient
		Try applying sample in different positions on the gel, with and without pre-running: some proteins might be altered only by a given pH; if sample has high molecular mass components, reduce the value of %T of the acrylamide gel; if protein aggregation due to high concn. of sample do not pre-run and set a low voltage, 100–200 V, for several hours to avoid concentrating effect of an established pH gradient at the beginning of the run; decrease protein load and use a more sensitive detection method; addition of detergent and/or urea is usually beneficial; if proteins are only sensitive to the ionic strength and/or the dielectric constant of the medium, increase carrier ampholyte concn. adding glycine or taurine; choice of denaturing conditions e.g. 8 M urea, detergents, and 2-mercaptoethanol, very often minimizes solubility problems

Continued

Troubleshooting

Table 1. Troubleshooting guide for proteins, *continued*

Symptom	Cause	Remedy
21. Plateau visible in the anodic and/or cathodic section of the gel during electrofocusing, no focusing proteins seen in that part of the gel	High concentration of salt in the system	Check correct amounts of APS and TEMED are used
22. Overheating of gel near sample application when beginning electrofocusing	High salt content in the sample	Reduce salt concentration by dialysis or gel filtration
23. Refractive line at pH 6.2 in the gel after focusing	Unincorporated polymers	Wash gel in 2 liters distilled water; change the water once and wash overnight
24. Curved protein zones in that portion of the gel which was at the top of the mold during polymerization	Too rapid polymerization	Decrease rate of polymerization by putting the mold into the freezer for 15 min before filling it with the gel solution, or place the solutions in a refrigerator for 15 min before casting the gel
25. Uneven protein distribution across a zone	Slot or sample application not perpendicular to running direction	Place slot or sample application pieces perfectly perpendicular to the running direction
26. No zones detected	Gel is focused with the wrong polarity	Mark polarity on the gel when removing it from the mold
27. Particulate material in the sample	May be using too low a ratio of solubilizing	Try using larger ratios of extraction

28. Reduction in sharpness and streaking of spots, and smearing at the alkaline end of the gel during sample loading	Interaction of nucleic acids with proteins and carrier ampholytes; can form a precipitate	In some cases nucleic acids can be removed by selective extraction or precipitation procedures [2–5]
29. Poor staining after SDS–PAGE; uneven staining with any buffer system; stained bands are being lost on destaining		Increase volume of staining solution to dilute out the SDS present; allow more time for staining step to permit the dye to penetrate fully; reduce destaining time or use a better fixative
30. Metallic sheen on gels after staining with Coomassie blue R-250	Solvent has been allowed to evaporate causing dye to dry on the gel at that point	Slight films of Coomassie blue sometimes observed on the gel surface after destaining can be removed by quick rinse in 50% methanol or by gently swabbing gel surface with methanol-soaked tissue paper
31. Blue or brown/black notches near gel borders after Coomassie blue or silver staining (respectively)	Fingerprints caused by gel handling without gloves	Always wear gloves
32. High background after silver-staining for proteins	Acrylic acid contamination in the acrylamide and/or bisacrylamide. Nucleic acids may also stain with silver stains	Highest quality reagents, including the purest (deionized) water, should always be used
33. Gels cracking during drying under vacuum	Vacuum released before the gel is properly dry	Take care to prevent such instances; thick and swollen gels require longer time to dry
34. Proteins left in gel after blotting	Protein probably precipitated in the gel	Add <0.01% SDS to transfer buffer

Continued

Table 1. Troubleshooting guide for proteins, *continued*

Symptom	Cause	Remedy
35. After blotting, proteins not in gel, and not present on blot	Protein lost	Ensure polarity of blotting used is correct (proteins should move towards positive pole)
36. After blotting no protein left in gel, but protein bands stain only faintly on blot. The front and back sides of the blot are equally stained	Protein moved through the filter, or protein was never present in gel; low binding capacity	Check sample protein concentration in gel; reduce detergent load in blotting buffer; check pH of blotting buffer and whether methanol added to transfer buffer
37. Staining intensity of protein in bands on nitrocellulose is weaker than expected	Not enough protein loaded; fraction of protein moved through the membrane	Determine protein concentration; also see no. 36
38. Protein bands appear fuzzy on the filter	Inefficient electrophoresis system; air bubbles between gel and nitrocellulose during transfer	Check electrophoresis system; remove air bubbles by passing glass pipette over membrane before blotting sandwich is assembled
39. Artifactual blackening of X-ray film during fluorography	Inadequate removal of DMSO	Ensure sufficient soaking in water before drying the gel

Table 2. Troubleshooting guide for nucleic acids

Symptom	Cause	Remedy
1. Gels do not polymerize, or polymerize unevenly	Traces of organic solvents, used to recrystallize the acrylamide or bisacrylamide, may be present and these inhibit polymerization; dissolved oxygen may be present in the gel mixture; faulty APS	Discard solutions and purchase and prepare fresh batch from pure reagents; ensure thorough degassing and efficient degassing apparatus; use freshly prepared APS, occasionally APS crystals are too old and a fresh supply must be used
2. Gels polymerize too quickly, especially with high concn. polyacrylamide gels		Reduce concn. of catalysts used; cool the mixture before adding the catalysts, hence slowing the rate of polymerization
3. Gels slide out of tubes after polymerization, especially with low concn. polyacrylamide gels		Cover the bottom of each tube with gauze or dialysis tubing secured with a rubber band; insert a disc of porous polythene into the bottom of each tube
4. No RNA detected in the gel after electrophoresis	Too little RNA loaded on to gel; contamination of the RNA with denatured proteins may cause the RNA to aggregate at the top of the gel; RNA may have been degraded to small fragments which pass straight through the gel; sample may have	Use a more conc. sample; try an alternative or additional deproteinizing step, i.e. use phenol/m-cresol instead of phenol/chloroform or use proteinase K to digest contaminating protein; use a more conc. gel or run for a shorter

Continued

Table 2. Troubleshooting guide for nucleic acids, *continued*

Symptom	Cause	Remedy
	floated off the gel during loading; the gel concn. may be too high	time and use bromophenol blue as a marker for the buffer front; if RNA is degraded, check starting material or extraction and purification steps; clean apparatus thoroughly and autoclave solutions to combat presence of ribonuclease; sample floating can be prevented by evaporating alcohol from the RNA during sample preparation or by adding more sucrose before loading; check recommended concn. used, vary gel concn.
5. RNA bands very broad or trailing	Inadequate sample preparation or loading	Check composition of electrophoresis buffer; check salt concn. in loading solution is not too high and the pH of the solution is as expected; load a smaller volume of RNA sample; use a lower conc. of RNA for loading, try an alternative deproteinizing procedure
6. Small amounts of RNA present in the gel are obscured by background 'noise' when scanning	Caused by impurities in the gel which absorb UV light or by dirt and dust introduced into the gel reagents, or adhered to the gels, or by air bubbles and	To remove high UV absorbance in the gels, wash gels in water for 1–2 h; remove impurities from sample before electrophoresis by re-precipitation

		or by washing RNA precipitates with ethanol and from the gels by recrystallization; filter all solutions and keep all apparatus clean; with very small amounts of RNA, use membrane filters to remove debris from solutions and exclude dust and air bubbles from cuvette when scanning
scratches on the cuvette or the gels		
7. Aggregation of RNA	Can be formed during RNA extraction at elevated temperatures or where RNA is dissolved at high concn.	Improve method of deproteinization or by denaturing the RNA before loading onto the gel; use denaturing conditions where previously used non-denaturing conditions
8. Trailing and broadening of zones	Increased flow of current results in the production of heat, which, if excessive, adversely causes the trailing and broadening of zones	Reduce current applied to gel
9. Undesirable pH ranges	Ionic strength is too low, which may seriously reduce the buffering capacity of the solution	Recirculate the buffer between the reservoirs or by changing it at intervals during the run
10. Difficulty in removing gels from tubes	Concentrated gels have tendency to be troublesome when trying to remove gels	Gels are looser if the SDS from the upper reservoir tank has migrated

Continued

Table 2. Troubleshooting guide for nucleic acids, *continued*

Symptom	Cause	Remedy
	from tubes	all the way through the gel. Pre-run gels for an hour or more before loading the samples
11. Artifactual bands in tube gels	Some RNA may accumulate at the interface between the two gels of a discontinuous gel	Use a continuous gradient of acrylamide or separate the RNA sample on two or three individual gels of different concn.
12. Faint bands or no bands, even when known DNA template is used	Insufficient or dirty template DNA; insufficient enzyme activity; poor annealing of primer to template; contamination of sequencing reaction with protein or salt; samples not denatured before loading on gel	Prepare new template DNA or use larger amount of DNA; use fresh or more enzyme; check that the primer sequence is correct for the template DNA, make sure the primer does not self-anneal or form hairpin structures; re-extract with phenol, excess salt can be removed by reprecipitating with ethanol and then washing the pellet with 70% ethanol before drying
13. Low band intensity at bottom or top of gel	Ratio of ddNTPs to dNTPs is too low or too high, respectively	Prepare fresh ddNTPs mixes or increase/decrease ratio of ddNTPs to dNTPs by a factor of 2–4
14. High background in each lane or a smear of uniform intensity along each lane	Contamination of template with RNA or polyethylene glycol	Prepare new template

15. Bands are fuzzy throughout the lanes	Dirty template DNA	Prepare new template
16. Bands at the same position in two or three lanes, occurring throughout the gel	DNA sample contains two different templates, generating overlapping sequences; primer hybridizing to a secondary site; priming occurring at nicks or gaps in DNA template or at contaminating DNA fragments	Prepare new template DNA from a single plaque or colony; increase stringency of annealing or make a new primer; use an end-labeled primer
17. Bands in the same position in all four lanes, occurring throughout the gel	Dirty template DNA; DNA template is nicked or contaminated with polyethylene glycol; insufficient enzyme activity	Prepare new template DNA; remove nicked DNA by acid–phenol extraction, remove excess polyethylene glycol by reprecipitating with ethanol; use a fresh preparation of enzyme or add more
18. Anomalous spacing of bands, missing bands at the same position in two or three lanes, occurring only at specific regions	Band compression, a newly synthesized DNA strand is forming secondary structure during gel electrophoresis, leading to anomalous migration	Increase temperature of gel electrophoresis; prepare the sequencing gel with 40% formamide
19. Bands in all four lanes, occurring at specific regions	Dissociation of enzyme from DNA template due to secondary structure in template	Use *Taq* DNA polymerase. Due to its high sequencing temperature, *Taq* DNA polymerase can proceed through secondary structures that would cause Klenow, reverse transcriptase or T7 DNA polymerase to dissociate from the template; perform a chase step to help eliminate false bands; increase the incubation temperatures for Klenow (42–55°C) and reverse transcriptase (up to 50°C)

Continued

Table 2. Troubleshooting guide for nucleic acids, *continued*

Symptom	Cause	Remedy
20. 'Smiling' in gel; bands which should be at the same level form a 'U' across the gel	Variation in temperature across the gel	Clamp a metal sheet, 2 mm thick aluminum is ideal, to the exposed gel plate
21. Poor resolution of oligodeoxynucleotides	Due to lack of or improper quantitation of the crude reaction mixture; insufficient migration as a result of too short a run time	Improve quantitation; run times should be estimated by the time it takes for a given species to migrate at least two-thirds the length of the gel and not judged by the arbitrary separation of the marker dyes
22. Broad smear of a higher average molecular weight than the desired product	Incomplete deprotection and unwanted chemical modifications	Re-check the chemical synthesis process
23. Trailing at edges of the bands	Caused by use of glycerol in loading buffer	Use Ficoll in loading buffer
24. Irregularities in polymerization of the gel	Loss of water during microwave heating of agarose solution; formation of agarose clumps on inside of flasks	Adjust for any loss of water after microwave heating; cool gradually to 35°C to prevent clumps forming; cool gel mold and sample well former at 20°C to permit gelling of agarose on the plates before polymerization of the acrylamide
25. Streaking of polysome bands, particularly at the margin of each lane	Overloading of gel; sample contains too much salt; surfaces of the gel plates are not clean	Gel sample into the well; thoroughly clean plates. Problems less severe when using a 3 mm thick gel with a 1 mm

		thick well since the sample is surrounded by gel which can to some extent compensate for overloading and salt effects. Also prevents sample from coming into contact with the gel plates
26. Streaked gel patterns with most eukaryotic polysomes	Association of membrane proteins with polysomes producing large aggregates may create considerable heterogeneity in particle size and account for the streaked pattern	Eukaryotic ribosomes from all sources can be separated as unfolded subunits in TBE-buffered gels. High salt washes and treatment with non-ionic detergents to remove non-ribosomal proteins may aid in resolution of polysomes from higher eukaryotic cells
27. Difficulty in drying gels down for autoradiography	Urea has not been sufficiently leached from the gel during fixation for autoradiography	Fixing for 15 min is usually sufficient; ensure close contact between the gel and the film
28. Bands are fuzzy in certain areas of gel	Poor contact of film with gel; wrinkle in dried gel	Ensure film is clamped tightly to gel; be careful to avoid wrinkles when drying gel
29. Entire film blank or nearly so	Essential component (enzyme, dNTPs, primer, radioactive precursor) omitted from reaction; saran wrap not removed from gel after drying	Check that the correct labeled dNTP was used; make a new batch of labeling and chain-termination mixtures; use a new batch of [^{35}S]dATP

Continued

111

Table 2. Troubleshooting guide for nucleic acids, *continued*

Symptom	Cause	Remedy
30. Black dots on autoradiograph	Precipitation of TBE or urea in gel mixtures	Prepare fresh gel mixtures using fresh components; ensure gel mixture is at RT when the gel is poured, otherwise urea may precipitate
31. Radioactivity remains in well	Samples not properly denatured	Transfer samples to microfuge tubes and denature by boiling for 5 min; chill samples to 0°C and load gel within 20 min

Chapter 8 **MANUFACTURERS AND SUPPLIERS**

Many of the larger companies have subsidiaries in other countries while most of the smaller companies market their own products or through agents. The name of a local supplier can be obtained by contacting the relevant company listed here. The numbers bracketed are area, freephone or freefax code numbers; international dialling codes have not been listed. UK (0800) and USA (800) freephone or freefax numbers can only be used in the corresponding countries.

Where (A) is found as a prefix before the company name, it indicates that this particular company supplies apparatus for gel electrophoresis; similarly, (C) indicates the sale of chemicals for gel electrophoresis. Some companies supply both as indicated. Some companies may not have any prefixes; these market items such as blotting membranes which do not essentially come under either of the categories.

AIP (A) **Amersham International Plc.,** Lincoln Place, Green End, Aylesbury, Buckinghamshire, HP20 2TP, UK.
Tel (0296) 395222, (0800) 515313.
Fax (0296) 85910.
2636 South Clearbrook Drive, Arlington Heights, IL 60005, USA.

ANA (A) **Anachem Ltd.,** Anachem House, 20 Charles Street, Luton, Bedfordshire, LU2 0EB, UK.
Tel (0582) 456666.
Fax (0582) 391768.

APP (A/C) **Appligene,** Pinetree Centre, Durham Road, Birtley, Chester-le-Street, Co. Durham, DH3 2TD, UK.
Tel (091) 492 0022.
Fax (091) 492 0617.
1177-C Quarry Lane, Pleasanton, CA 94566, USA.
Tel (510) 462 2232.

Fax (510) 462 6247.

BDH (A/C) **BDH Laboratories,** Merck Ltd., Merck House, Poole, Dorset, BH15 1TD, UK.
Tel (0202) 669700, (0800) 223344.
Fax (0202) 665599.

BHM (C) **Boehringer Mannheim,** Bell Lane, Lewes, East Sussex, BN7 1LG, UK.
Tel (0273) 480444, (0800) 521578.
9115 Hague Road, PO Box 50414, Indianapolis, IN 46250-0414, USA.
Tel (800) 262 1640.
Fax (317) 576 2754.

BML (A) **Biometra Ltd.,** Whatman House, St. Leonard's Road, 20/20 Maidstone, Kent, ME16 0LS, UK.
Tel (0622) 678872.
Fax (0622) 752774.

BRL (A/C) **Bio-Rad Laboratories Ltd.,** Bio-Rad House, Maylands Avenue, Hemel Hempstead, Hertfordshire, HP2 7TD, UK.
Tel (0442) 232552, (0800) 181134.
Fax (0442) 259118.

2000 Alfred Nobel Drive, Hercules, CA 94547, USA.
Tel (516) 756 2575.
Fax (516) 756 2594.

BSS (A) **Bioscience Services,** Schiehallion, Woburn Close, Cramlington, Northumberland, NE23 9QP, UK.
Tel and Fax (0670) 736590.

DNC (A) **Dionex Corporation,** Albany Court, Camberley, Surrey, GU15 2PL, UK.
Tel (0276) 691722.
Fax (0276) 691837.
1228 Titan Way, PO Box 3603, Sunnyvale, CA 94088-3603, USA.
Tel (408) 737 0700.
Fax (408) 730 9403.

ECA (A) **E-C Apparatus Corporation,** for UK see LSL; 3831 Tyrone Boulevard North, St. Petersburg, FL 33709, USA.
Tel (813) 344 1644.
Fax (813) 343 5730.

EKL (A/C) **Eastman Kodak Ltd.,** PO Box 33, Swallowdale Lane, Hemel Hempstead, Hertfordshire, HP2 7EU, UK.
Tel (0442) 42281.
Fax (0442) 230367.
25 Science Park, New Haven, CT 06511, USA.
Tel (203) 786 5600, (800) 225 5352.
Fax (203) 624 3143, (800) 879 4979.

FIL (A/C) **Flowgen Instruments Ltd.,** Broad Oak Enterprise Village, Broad Oak Road, Sittingbourne, Kent, ME9 8AQ, UK.
Tel (0795) 429737.
Fax (0795) 471185.

FSE (A/C) **Fisons Scientific Equipment,** Bishop Meadow Road, Loughborough, Leicestershire, LE11 0RG, UK.
Tel. (0509) 231166.
Fax. (0509) 231893.

GBL (A/C) **Gibco–BRL,** Life Technologies, Unit 4, Cowley Mill Trading Estate, Longbridge Way, Uxbridge, Middlesex, UB8 2YG, UK.

Tel (0895) 36355, (0800) 838380.
Fax (0895) 53159.
8400 Helgerman Court, Gaithersburg, MD 20877, USA.
Tel (301) 840 800.

GKI (A) **Glyko Inc.,** for UK see MPC; 81 Digital Drive, Novato, CA 94949, USA.
Tel (415) 382 7889.
Fax (415) 382 6653.

GRI (A) **Genetic Research Instrumentation Ltd,** Gene House, Dunmow Road, Felsted, Dunmow, Essex, CM6 3LD, UK.
Tel (0371) 821082.
Fax (0371) 820131.

HLL (A) **Helena Laboratories,** Seventh Avenue, Team Valley Trading Estate, Gateshead, Tyne and Wear, NE11 0LN, UK.
Tel (091) 487 8855.
Fax (091) 491 0602.
1530 Lindbergh Drive, PO Box 752, Beaumont, TX 77704-0752, USA.

Tel (409) 842 3714, (800) 231 5663.
Fax (409) 842 6241.

HSI   (A/C) **Hoefer Scientific Instruments,** Unit 12, Croft Road Workshops, Croft Road, Newcastle- under-Lyme, Staffordshire, ST5 0TW, UK.
Tel (0782) 617317.
Fax (0782) 617346.
654 Minnesota Street, PO Box 77387, San Francisco, CA 94107, USA.
Tel (415) 282 2307, (800) 227 4750.
Fax (415) 821 1081.

IBI   (A/C) **International Biotechnologies, Inc.,** 36 Clifton Road, Cambridge, CB1 4ZR, UK.
Tel (0223) 242813.
Fax (0223) 243036.
For USA, see EKL.

ICN   (C) **ICN Flow Biomedicals,** Eagle House, Penegrine Business Park, Gomm Road, High Wycombe, Buckinghamshire, HP13 7DL, UK.
Tel (0494) 443826.
Fax (0494) 473162.

3300 Hyland Avenue, Costa Mesa, CA 92626, USA.
Tel (800) 545 0530.
Fax (800) 334 6999.

JNP   **Janssen Pharmaceutical,** Hyde Park House, Cartwright Street, Newton, Hyde, Cheshire, SK14 4EH, UK.
Tel (0613) 679277.
Fax (0613) 678165.

JSL   (A) **Jencons (Scientific) Ltd.,** Cherrycourt Way Industrial Estate, Stanbridge Road, Leighton Buzzard, LU7 8UA, UK.
Tel (0525) 372010.
Fax (0525) 379547.

LSI   (A) **Life Sciences International (Europe) Ltd.,** Chadwick Road, Astmoor, Runcorn, Cheshire, WA7 1PR, UK.
Tel (0928) 566611.
Fax (0928) 565845.

MDI   (A) **Molecular Dynamics Inc.,** 4 Chaucer Business Park, Kemsing, Sevenoaks, Kent, TN15 6PL, UK.

Tel (0732) 62565.
Fax (0732) 63422.
880 E. Arques Avenue, Sunnyvale, CA 94086, USA.
Tel (408) 773 1222.
Fax (408) 773 8343.

MPC (A/C) **Millipore Corporation,** The Boulevard, Blackmoor Lane, Watford, Hertfordshire, WD1 8YW, UK.
Tel (0923) 816375.
Fax (0923) 818297.
PO Box 255, Bedford, MA 01730, USA.
Tel (617) 275 9200, (800) 225 1380.

NBL (A/C) **Northumbria Biologicals Ltd.,** Nelson Industrial Estate, Cramlington, Northumberland, NE23 9BL, UK.
Tel (0670) 732992.
Fax (0670) 732537.

NEN **NEN Products,** Du Pont Ltd., Wedgwood Way, Stevenage, Hertfordshire, SG1 4QN, UK.
Tel (0438) 734026.
Fax (0438) 734379.

PMB (A/C) **Pharmacia Biosystems,** 23 Grosvenor Road, St. Albans, Hertfordshire, AL1 3AW, UK.
Tel (0727) 814000.
Fax (0727) 814001.
800 Centennial Avenue, PO Box 1327, Piscataway, NJ 08855 1327, USA.
Tel (201) 457 8000.
Fax (201) 457 0557.

PML (A/C) **Promega Ltd.,** Delta House, Enterprise Road, Chilworth Research Centre, Southampton, SO1 7NS. UK.
Tel (0703) 760225, (0800) 378994.
Fax (0703) 767014, (0800) 181037.
Madison, Wisconsin, USA.
Tel (800) 356 9526.
Fax (608) 273 6967.

RSL (A/C) **Rotec Scientific Ltd.** (Carl Roth), 10 Bridgeturn Avenue, Old Wolverton, Milton Keynes, MK12 5QL, UK.
Tel (0908) 223399.
Fax (0908) 223000.

S&S **Schleicher & Schuell,** 10 Optical Avenue, Keene, NH 03431, USA.
Tel (603) 352 3810.
Fax (603) 357 3627.

SCC (A/C) **Sigma Chemical Corporation,** Fancy Road, Poole, Dorset, BH17 7BR, UK.
Tel (0202) 733114, (0800) 447788, Overseas call reverse charge (0202) 733114.
Fax (0202) 715460.
PO Box 14508, 3500 DeKalb Street, St. Louis, MO 63178, USA.
Tel (800) 848 7791, Overseas call collect (314) 771 5765.

SGB (C) **Serva GmBH,** 50 A&S Drive, Paramus, NJ 07652, USA.
Tel (800) 645 3412.
Fax (201) 967 8858.

SPL (A) **Scie-Plas Ltd.,** Unit 2, Cottage Leap, Butlers Leap, Rugby, CV21 3XP, UK.
Tel (0788) 551655.
Fax (0788) 551411.

SSI (A) **Shandon Southern Instruments,** for UK see LSL; 515 Broad Street, Drawer 43, Sewickley, PA 15143 0043, USA.
Tel (412) 741 8400.

STG (A/C) **Stratagene,** Cambridge Innovation Centre, Cambridge Science Park, Milton Road, Cambridge, CB4 4GF, UK.
Tel (0223) 420955, (0800) 585370.
Fax (0223) 420234.
11099 North Torrey Pines Road, La Jolla, CA 92037, USA.
Tel (619) 535 5400, (800) 424 5444.
Fax (619) 535 0045.

USB **United States Biochemical,** PO Box 22400, Cleveland, OH 44122, USA.
Tel (216) 765 5000, (800) 321 9322.
Fax (216) 464 5075, (800) 535 0898.

USL (A) **Uniscience Ltd.,** Wildmere Road, Banbury, Oxon, OX16 7JU, UK.
Tel (0295) 272270.
Fax (0295) 272243.

UVP (A) **Ultra-Violet Products Ltd.,** Science Park, Milton Road, Cambridge, CB4 4FH, UK.
Tel (0223) 420022.
Fax (0223) 420561.

VAH (A) **V. A. Howe & Co Ltd.,** Beaumont Close, Banbury, Oxon, OX16 7RG, UK.
Tel (0295) 252666.
Fax (0295) 268096.

VBL (A) **Vilber Lourmat,** for UK see USL; B. P. 66 Torcy, Z. I. Sud - 77202, Marne La Vallée, Cedex 1, France.
Tel (1) 60 06 07 71.
Fax (1) 64 80 48 59.

REFERENCES

Chapter 4

1. Weber, K. and Osborn, M. (1969) *J. Biol. Chem.* **244,** 4406.
2. Laemmli, U.K. (1970) *Nature* **227,** 680.
3. O'Farrell, P.Z., Goodman, H.M. and O'Farrell, P.H. (1977) *Cell* **12,** 1133.
4. Swank, R.W. and Munkres, K.D. (1971) *Anal. Biochem.* **39,** 462.
5. Hashimoto, F., Horigome, T., Kanbayashi, M., Yoshida, K. and Sugano, H. (1983) *Anal. Biochem.* **129,** 192.
6. McLellan, T. (1982) *Anal. Biochem.* **126,** 94.
7. Reisfeld, R.A., Lewis, V.J. and Williams, D.E. (1962) *Nature* **195,** 281.
8. Williams, D. E. and Reisfeld, R.A. (1964) *Ann. NY Acad. Sci.* **121,** 373.
9. Davis, B.J. (1964) *Ann. NY Acad. Sci.* **121,** 404.
10. Righetti, P. G. (1983) *Isoelectric Focusing: Theory, Methodology and Applications.* Elsevier, Amsterdam.
11. LKB Application Note 324 (1984).
12. Görg, A., Fawcett, J.S. and Chrambach, A. (1988) in *Advances in Electrophoresis* (A. Chrambach, M.J. Dunn, and B.J. Radola, eds), Vol. 2, p. 1. VCH, Weinheim.
13. Gianazza, E., Astrua-Testori, S. and Righetti, P.G. (1985) *Electrophoresis* **6,** 113.
14. Bjellqvist, B., Ek, K., Righetti, P.G., Gianazza, E., Görg, A., Postel, W. and Westermeier, R. (1982) *J. Biochem. Biophys. Methods* **6** 317.
15. LKB Application Note 321.
16. O'Farrell, P.H. (1975) *J. Biol. Chem.* **250,** 4007.
17. Chambers, J.A.A., Hinkelammert, K., degli Innocenti, F. and Russo, V.E.A. (1985) *Electrophoresis* **6,** 339.
18. MacGillivray, A.J. and Rickwood, D. (1985) *Biochem. J.* **229,** 771.
19. Manabe, T., Tachi, K., Kojima, K. and Okuyama, T. (1979) *J. Biochem.* **85,** 649.
20. Kaltschmidt, E. and Wittman, H.G. (1970) *Anal. Biochem.* **36,** 401.
21. Howard, G.A. and Traut, R.R. (1973) *FEBS Lett.* **29,** 177.
22. Reisfeld, R.A., Lewis, V.J. and Williams, D.E. (1962) *Nature* **195,** 281.

23. Thomas, J. and Kornberg, R. (1975) *Proc. Natl Acad. Sci. USA* **72**, 2626.

24. Hoffman, P. and Chalkey, R. (1976) *Anal. Biochem.* **76**, 539.

25. Pipkin, J., Anson, J.F., Hinson, W.G., Burns, E.R. and Wolff, G.L. (1985) *Electrophoresis* **6**, 306.

26. Orrick, L., Ohlson, M. and Busch, H. (1973) *Proc. Natl Acad. Sci. USA.* **70**, 1316.

27. Ames, G.F.L. and Nikaido, K. (1976) *Biochemistry* **15**, 616.

28. Anderson, N. G. and Anderson, N. L. (1978) *Anal. Biochem.* **85**, 331.

29. Anderson, N. L. and Anderson, N. G. (1978) *Anal. Biochem.* **85**, 341.

30. Dunbar, B. S. (1987) *Two-dimensional Gel Electrophoresis and Immunological Techniques.* Plenum Press, New York.

31. Horst, M. N., Mahaboob, S., Basha, M., Baumbach, G. A., Mansfield, E. H., and Roberts, E. M. (1980) *Anal. Biochem.* **102**, 399.

32. Wilson, D., Hall, M. E., Stone, G. C., and Rubin, R. W. (1977) *Anal. Biochem.* **83**, 33.

33. Garrels, J. I. (1979) *J. Biol. Chem.* **254**, 7961.

34. Horikawa, S. and Ogawara, H. (1979) *Anal. Biochem.* **97**, 116.

35. Rottem, S., Stein, O. and Razin, S. (1968) *Arch. Biochem. Biophys.* **125**, 46.

36. Loach, P.A., Sekura, D.L., Hadsell, R.M. and Stemer, A. (1970) *Biochemistry* **9**, 724.

37. Holloway, P.W. (1973) *Anal. Biochem.* **53**, 304.

38. Fasman, G. D. (ed.) (1976) *Handbook of Biochemistry and Molecular Biology, Proteins,* 3rd Edn., Vol. II. CRC Press, Cleveland, Ohio, USA.

39. Malamud, D. and Drysdale, J. W. (1978) *Anal. Biochem.* **86**, 620.

40. Dautrevaux, M., Boulanger, Y., Han, K. and Biserte, G. (1969) *Eur. J. Biochem.* **11**, 267.

41. Lambin, P. C. (1978) *Anal. Biochem.* **85**, 114.

42. Weber, K. and Osborn, M. (1969) *J. Biol. Chem.* **244**, 4406.

43. Cowman, M.K., Slahetka, M.F., Hittner, D.M., Kim, J., Forino, M. and Gadelrad, G. (1984) *Biochem. J.* **221**, 707.

44. Laskey, R. A. and Mills, A. D. (1975) *Eur. J. Biochem.* **56**, 335.

45. Towbin, H., Staehelin, T. and Gordon, J. (1979) *Proc. Natl Acad. Sci. USA* **76**, 4350.

46. Blakesly, R. W. and Boezi, J. A. (1977) *Anal. Biochem.* **82**, 580.

47. Vesterberg, O. (1972) *Biochim. Biophys. Acta* **257**, 11.

48. Righetti, P. G. and Drysdale, J. W. (1974) *J. Chromatogr.* **98**, 271.

49. Neuhoff, V., Stamm, R. and Eibl, H. (1985) *Electrophoresis* **6**, 427.

121

50. Lee, C., Levin, A. and Branton, D. (1987) *Anal. Biochem.* **166,** 308.

51. Allen, R.E., Masak, K.C. and McAllister, P.K. (1980) *Anal. Biochem.* **104,** 494.

52. Laskey, R. A. (1980) in *Methods in Enzymology.* (L. Grossman and K. Moldave, eds), Vol. 65, p. 363. Academic Press, New York.

53. Richtie, R. F. and Smith, R. (1976) *Clin. Chem.* **22,** 497.

54. Hebert, J. P. and Strobbel, B. (1974) LKB Application Note 151.

55. Arnaud, P., Wilson, G. B., Koistinen, J. and Fudenberg, H.H. (1977) *J. Immunol. Methods* **16,** 221.

56. Merril, C. R., Goldman, D., Sedman, S. A. and Ebert, M. H. (1981) *Science* **211,** 1438.

57. Godolphin, W. J. and Stinson, R. A. (1974) *Clin. Chim. Acta* **56,** 97.

58. Harris, H. and Hopkinson, D. A. (1976) *Handbook of Enzyme Electrophoresis in Human Genetics.* Elsevier, Amsterdam.

59. Sun, S.M. and Hall, T.C. (1974) *Anal. Biochem.* **61,** 237.

60. Bossard, H.F. and Datyner, A. (1977) *Anal. Biochem.* **82,** 327.

61. Schägger, H., Aquila, H. and van Jagow, G. (1988) *Anal. Biochem.* **173,** 201.

62. Stephens, R.E. (1975) *Anal. Biochem.* **65,** 369.

63. Schetters, H. and McLeod, B. (1979) *Anal. Biochem.* **98,** 329.

64. Tijssen, P. and Kurstak, E. (1979) *Anal. Biochem.* **99,** 97.

65. Eng, P.R. and Parker, C.O. (1974) *Anal. Biochem.* **59,** 323.

66. Ragland, W.L., Pace, J.L. and Kemper, D.L. (1974) *Anal. Biochem.* **59,** 24.

67. Douglas, S.A., La Marca, M.E. and Mets, L.J. (1978) in *Electrophoresis '78* (N. Catsimpoolas, ed.), Vol. 2, p. 155. Elsevier, Amsterdam.

68. Ragland, W.L., Benton, T.L., Pace, J.L., Beach, F.G. and Wade, A.E. (1978) in *Electrophoresis '78* (N. Catsimpoolas, ed.), Vol. 2, p. 217. Elsevier, Amsterdam.

69. Yamamoto, K., Okamoto, Y. and Sekine, T. (1978) *Anal. Biochem.* **84,** 313.

70. Burger, B.O., White, F.C., Pace, J.L., Kemper, D.L. and Ragland, W.L. (1976) *Anal. Biochem.* **70,** 327.

71. Urwin, V.E. and Jackson, P. (1991) *Anal. Biochem.* **195,** 30.

72. Weidekamm, E., Wallach, D.F.H. and Flückiger, R. (1973) *Anal. Biochem.* **54,** 102.

73. Hartman, B.K. and Udenfriend, S. (1969) *Anal. Biochem.* **30,** 391.

74. Harowitz, P.M. and Bowman, S. (1987) *Anal. Biochem.* **165,** 430.

75. Jackowski, G. and Liew, C.C. (1980) *Anal. Biochem.* **102,** 34.

76. Carson, S.D. (1977) *Anal. Biochem.* **78,** 428.

77. Liebowitz, M.J. and Wang, R.W. (1984) *Anal. Biochem.* **137**, 161.

78. Andrews, A.T. (1986) *Electrophoresis: Theory, Techniques and Biochemical and Clinical Applications,* p. 29. Clarendon Press, Oxford.

79. Wallace, R.W., Yu, P.H., Dieckart, J.P. and Dieckart, J.W. (1974) *Anal. Biochem.* **61**, 86.

80. Nelles, L.P. and Bamburg, J.R. (1976) *Anal. Biochem.* **73**, 522.

81. Tagi, T., Kubo, K. and Isemura, T. (1977) *Anal. Biochem.* **79**, 104.

82. Higgins, R.C. and Dahmus, M.E. (1979) *Anal. Biochem.* **93**, 257.

83. Mardian, J.K.W. and Isenburg, I. (1978) *Anal. Biochem.* **91**, 1.

84. Olden. K. and Yamada, K.M. (1977) *Anal. Biochem.* **78**, 483.

85. van Raamsdonk, W., Pool, C.W. and Heyting, C. (1977) *J. Immunol. Methods* **17**, 337.

86. Parish, R.W., Schmidlin, S. and Parish, C.R. (1978) *FEBS Lett.* **95**, 366.

87. Stumph, W.E., Elgin, S.C.R. and Hood, L. (1974) *J. Immunol.* **113**, 1752.

88. Groschel-Stewart, U., Schreiber, J., Mahlmeister, C. and Weber, K. (1976) *Histochemistry* **46**, 229.

89. Burridge, K. (1978) in *Methods in Enzymology* (V. Ginsburg, ed.), Vol. 50, p. 54. Academic Press, New York.

90. Kasamatsu, H. and Flory, P.J. (1978) *Virology* **86**, 344.

91. Adair, W.S., Jurivich, D. and Goodenough, U.W. (1978) *J. Cell Biol.* **79**, 281.

92. Bigelis, R. and Burridge, K. (1978) *Biochem. Biophys. Res. Commun.* **82**, 322.

93. Saltzgaber-Müller, J. and Schatz, G. (1978) *J. Biol. Chem.* **253**, 305.

94. West, C.M. and McMahon, D. (1977) *J. Cell Biol.* **74**, 264.

95. Gander, J.E. (1984) in *Methods in Enzymology* (W.B. Jakoby, ed.), Vol. 104, p. 447. Academic Press, New York.

96. Carson, S.D. (1977) *Anal. Biochem.* **78**, 428.

97. Gahmberg, C.G. (1978) in *Methods in Enzymology* (V. Ginsburg, ed.), Vol. 50, p. 204. Academic Press, New York.

98. Wallenfels, B. (1979) *Proc. Natl Acad. Sci. USA.* **76**, 3223.

99. Gregg, J.H. and Karp, G.C. (1978) *Exp. Cell Res.* **112**, 31.

100. Bradshaw, J.P. and White, P.A. (1985) *Biosci. Rep.* **5**, 229.

101. Taylor, T. and Weintraub, B.D. (1985) *Endocrinology.* **116**, 1968.

102. Avigad, G. (1978) *Anal. Biochem.* **86**, 443.

103. Moroi, M. and Jung, S.M. (1984) *Biochem. Biophys. Acta* **798**, 295.

104. Fairbanks, G., Steck, T.L. and Wallach, D.L.H. (1971) *Biochemistry* **10**, 2026.

105. Wardi, A.H. and Michos, G.A. (1972) *Anal. Biochem.* **49**, 607.

References

106. Furlan, M., Perret, B.A. and Beck, E.A. (1979) *Anal. Biochem.* **96,** 208.

107. Dubray, G. and Bezard, G. (1982) *Anal. Biochem.* **119,** 325.

108. Dupuis, G. and Doucet, J.P. (1981) *Biochim. Biophys. Acta* **669,** 171.

109. Koch, G.L.E. and Smith, M.J. (1982) *Eur. J. Biochem.* **128,** 107.

110. King, L.E. and Morrison, M. (1976) *Anal. Biochem.* **71,** 223.

111. Goldman, D., Merril, C.R. and Ebert, M.H. (1980) *Clin. Chem.* **26,** 1317.

112. Tsai, C.M. and Frasch, C.E. (1982) *Anal. Biochem.* **119,** 115.

113. Ressler, N., Springate, R. and Kaufman, J. (1961) *J. Chromatogr.* **6,** 409.

114. Prat, J.P., Lamy, J.N. and Weill, J.D. (1969) *Bull. Soc. Chim. Biol.* **51,** 1367.

115. Debruyne, I. (1983) *Anal. Biochem.* **133,** 110.

116. Cutting, J.A. (1984) in *Methods in Enzymology* (W.B. Jakoby, ed.), Vol. 104, p. 451. Academic Press, New York.

117. Satoh, K. and Busch, H. (1981) *Cell Biol. Int. Rep.* **5,** 857.

118. Green, M.R., Pastewka, J.V. and Peacock, A.C. (1973) *Anal. Biochem.* **56,** 43.

119. Gabriel, O. (1971) in *Methods in Enzymology* (S.P. Colowick and N.O. Kaplan, eds), Vol. 22, p. 578. Academic Press, New York.

120. Siciliano, M.J. and Shaw, C.R. (1976) in *Chromatographic and Electrophoretic Techniques* (I. Smith, ed.), Vol. 2, p. 185. William Heinemann Medical Books Ltd., London.

121. Harris, H. and Hopkinson, D.A. (1976) *Handbook of Enzyme Electrophoresis in Human Genetics*. North-Holland, Amsterdam.

122. Shaw, C.R. and Prasad, R. (1980) *Biochem. Genet.* **4,** 297.

123. Harris, H. and Hopkinson, D.A. (1978) *Handbook of Enzyme Electrophoresis in Human Genetics*. North-Holland, Amsterdam.

124. Harris, H. and Hopkinson, D.A. (1977) *Handbook of Enzyme Electrophoresis in Human Genetics*. North-Holland, Amsterdam.

125. Abraham, D.G. and Cooper, A.J.L. (1991) *Anal. Biochem.* **197,** 421.

126. Blank, A., Suigiyama, R.H. and Dekker, C.A. (1982) *Anal. Biochem.* **130,** 267.

127. Mathew, R. and Rao, K.K. (1992) *Anal. Biochem.* **206,** 50.

128. Grenier, J. and Asselin, A. (1993) *Anal. Biochem.* **212,** 301.

129. Steiner, B. and Cruce, D. (1992) *Anal. Biochem.* **200,** 405.

130. Mittler, R. and Zilinskas, B.A. (1993) *Anal. Biochem.* **212,** 540.

131. Akyama, S. (1990) *Electrophoresis* **11,** 509.

132. Queiroz-Clapet, C. and Meunier, J.-C. (1993) *Anal. Biochem.* **209,** 228.

133. Gershoni, J.M. (1987) in *Advances in Electrophoresis* (A. Chrambach, M.J. Dunn and B.J. Radola, eds), Vol. 1, p. 141. VCH, Weinheim.

134. Bjerrum, O.J. and Schafer-Nielsen, C. (1986) in *Electrophoresis* (M.J. Dunn, ed.), p. 315, VCH, Weinheim.

135. Dunn, S.D. (1986) *Anal. Biochem.* **157**, 144.

136. In *Protein Blotting, a Guide to Transfer and Detection.* Bio-Rad Laboratories, Bulletin 1721. Bio-Rad Laboratories (1993)

137. Glass, W.F., Briggs, R.C. and Hnilica, L.S. (1981) *Anal. Biochem.* **115**, 219.

138. Keren, Z., Berke, G. and Gershoni, J.M. (1986) *Anal. Biochem.* **155**, 182.

139. Bradbury, W.C., Mills, S.D., Preston, M.A., Barton, L.J. and Penner, J.L. (1984) *Anal. Biochem.* **137**, 129.

140. Roberts, P.L. (1985) *Anal. Biochem.* **147**, 521.

141. Blake, M.S., Johnson, K.H., Russel-Jones, G.J. and Gotschlich, E.C. (1984) *Anal. Biochem.* **136**, 175.

142. Towbin, H. and Gordon, J. (1984) *J. Immunol. Methods* **72**, 313.

143. Weigele, M., De Bernado, S., Leimgruber, W., Cleeland, R. and Grunber, E. (1973) *Biochem. Biophys. Res. Commun.* **54**, 899.

144. Hsu, Y. (1984) *Anal. Biochem.* **142**, 221.

145. Renart, J. and Sandoval, I.V. (1984) in *Methods in Enzymology* (W.B. Jakoby, ed.), Vol. 104, p. 455. Academic Press, New York.

146. Kuones, D.R., Roberts, P.J. and Cottingham, I.R. (1986) *Anal. Biochem.* **152**, 221.

147. Walton, K.E., Styer, D. and Gruenstein, E. (1979) *J. Biol. Chem.* **254**, 795.

148. Cooper, P.C. and Burgess, A.W. (1982) *Anal. Biochem.* **126**, 301.

149. Davidson, J.B. and Case, A. (1982) *Science* **215**, 1398.

150. Burbeck, S. (1983) *Electrophoresis* **4**, 127.

151. Bonner, W.M. (1984) in *Methods in Enzymology* (W.B. Jakoby, ed.), Vol. 104, p. 461. Academic Press, New York.

152. Skinner, K. and Griswold, M.D. (1983) *Biochem. J.* **209**, 281.

153. Chamberlain, J.P. (1979) *Anal. Biochem.* **98**, 132.

154. Heegard, N.H.H., Hebsgaard, K.P. and Bjerrum, O.J. (1984) *Electrophoresis* **5**, 230.

155. McConkey, E.H. and Anderson, C. (1984) *Electrophoresis* **5**, 230.

156. Laskey, R.A. (1981) *Amersham Research News* No. 23.

157. Laskey, R.A. and Mills, A.D. (1977) *FEBS Lett.* **82**, 314.

158. Bonner, W.M. (1983) in *Methods in Enzymology* (S. Fleischer and B. Fleischer, eds), Vol. 96, p. 215. Academic Press, New York.

125

159. Harding, C.R. and Scott, I.R. (1983) *Anal. Biochem.* **129,** 371.
160. Latter, G.I., Burbeck, S., Fleming, S. and Leavitt, J. (1984) *Clin. Chem.* **30,** 1925.
161. Christopher, A.R., Nagpal, M.L., Carrol, A.R. and Brown, J.C. (1978) *Anal. Biochem.* **85,** 404.
162. Zapolski, E.J., Gersten, D.M. and Ledley, R.S. (1982) *Anal. Biochem.* **123,** 325.
163. Rickwood, D., Patel, D. and Billington, D. (1993) in *Biochemistry Labfax* (J.A.A. Chambers and D. Rickwood, eds), BIOS Scientific Publishers Ltd., Oxford.
164. Blank, A., Silber, J.R., Thelen, M.P. and Dekker, C.A. (1983) *Anal. Biochem.* **135,** 423.
165. Manrow, R.E. and Dottin, R.P. (1982) *Anal. Biochem.* **120,** 181.
166. Manrow, R.E. and Dottin, R.P. (1980) *Proc. Natl. Acad. Sci. USA* **77,** 730.
167. Hames, B.D. (1990) in *Gel Electrophoresis of Proteins, a Practical Approach* (B.D. Hames and D. Rickwood, eds), 2nd Edn, p. 1. IRL Press, Oxford.
168. Shackelford, D.A. and Zivin, J.A. (1993) *Anal. Biochem.* **211,** 131.
169. Weber, K. and Kuter, D.J. (1971) *J. Biol. Chem.* **246,** 4505.
170. Hager, D.A. and Burgess, R.R. (1980) *Anal. Biochem.* **109,** 76.
171. Konigsberg, W.H. and Henderson, H. (1983) in *Methods in Enzymology* (C.H.W. Hirs and S.N. Timasheff, eds), Vol. 91, p. 254. Academic Press, New York.

Chapter 5

1. Gross, K., Probst, E., Schaffner, N. and Birnstiel, M.L. (1976) *Cell* **8,** 455.
2. Bailey, J.M. and Davidson, N. (1976) *Anal. Biochem.* **70,** 75.
3. De Wachter, R. and Fiers, W. (1972) *Anal. Biochem.* **49,** 184.
4. Frisby, D. P., Newton, C., Carey, N.H., Fellner, P., Newman, J.F.E., Harris, T.J.R. and Brown, F. (1976) *Virology* **71,** 379.
5. Kennedy, S.I.T. (1976) *J. Mol. Biol.* **108,** 491.
6. Pedersen, F.S. and Haseltine, W.A. (1980) in *Methods in Enzymology* (L. Grossman and K. Moldave, eds), Vol. 65, p. 680. Academic Press, New York.
7. Vigne, R. and Jordon, B.R. (1971) *Biochimie* **53,** 981.
8. Ikemura, T. and Dahlberg, J.E. (1973) *J. Biol. Chem.* **248,** 5024.
9. Varricchio, F. and Ernst, H.J. (1975) *Anal. Biochem.* **68,** 485.
10. Burckhardt, J. and Birnstiel, M.L. (1978) *J. Mol. Biol.* **118,** 61.
11. Fischer, S.G. and Lerman, L.S. (1979) *Cell* **16,** 191.
12. Fischer, S.G. and Lerman, L.S. (1980) in *Methods in*

Enzymology (R. Wu, ed.), Vol. 68, p. 183. Academic Press, New York.

13. Sandeen, G., Wood, W.I. and Felsenfeld, G. (1980) *Nucleic Acids Res.* **8,** 3757.
14. Dahlberg, A.E., Lund, E. and Kjeldgaard, N.O. (1973) *J. Mol. Biol.* **78,** 627.
15. Dahlberg, A.E. (1974) *J. Biol. Chem.* **249,** 7673.
16. Tokomatsu, H., Strycharz, W. and Dahlberg, A.E. (1981) *J. Mol. Biol.* **152,** 397.
17. Peacock, A.C. and Dingham, C.W. (1968) *Biochemistry* **7,** 668.
18. Dahlberg, A.E., Dahlberg, J.E., Lund, E., Tokimatsu, H., Rabson, A.B., Calvert, P.C., Reynolds, F. and Zahalak, M. (1978) *Proc. Natl Acad. Sci. USA.* **75,** 3598.
19. Brown, T.A. (ed.) (1991) *Molecular Biology Labfax.* BIOS Scientific Publishers Ltd., Oxford.
20. Laskey, R.A. and Mills, A.D. (1975) *Eur. J. Biochem.* **56,** 355.
21. Bonner, W.M. and Laskey, R.A. (1974) *Eur. J. Biochem.* **46,** 83.
22. Probst, E., Dressman, A. and Birnstiel, M.L. (1979) *J. Mol. Biol.* **135,** 709.

23. Chamberlain, J.P. (1979) *Anal. Biochem.* **98,** 132.

Chapter 6

1. Jackson, P. (1990) *Biochem J.* **270,** 705.
2. Jackson, P. (1991) *Anal. Biochem.* **196,** 238.

Chapter 7

1. Righetti, P.G. and Macelloni, C. (1982) *J. Biochem. Biophys. Methods* **6,** 1.
2. Johns, E.W. (1976) *Subcellular Components: Preparation and Fractionation* (G.D. Birnie, ed.), p. 202. Butterworths, London.
3. Kruh, J., Schapira, G., Lareau, J. and Dreyfus, J.C. (1964) *Biochem. Biophys. Acta* **87,** 669.
4. Adamietz, P. and Hiltz, H. (1976) *Hoppe-Seylers Z. Physiol. Chem.* **357,** 527.
5. Sinclair, J.H. and Rickwood, D. (1985) *Biochem. J.* **229,** 771.

FURTHER READING

1 Books

Babskii, V.G., Zhukov, M.Y. and Yudovich, V.I. (1989) *Mathematical Theory of Electrophoresis*. Consultants Bureau, New York.

Brown, T.A. (ed.) (1991) *Molecular Biology Labfax*. BIOS Scientific Publishers Ltd., Oxford.

Celis, J.E. and Bravo, R. (1984) *Two-Dimensional Gel Electrophoresis of Proteins*. Academic Press, Orlando.

Chambers, J.A.A. and Rickwood, D. (eds.) (1993) *Biochemistry Labfax*. BIOS Scientific Publishers Ltd., Oxford.

Dunbar, B.S. (1987) *Two-Dimensional Electrophoresis and Immunological Techniques*. Plenum Press, New York.

Dunn, M.J. (1993) *Gel Electrophoresis: Proteins*. BIOS Scientific Publishers Ltd., Oxford.

Hames, B.D. and Rickwood, D. (eds) (1990) *Gel Electrophoresis of Proteins, A Practical Approach*, 2nd Edn. IRL Press, Oxford.

Mosher, R.A., Saville, D.A. and Thosmann, W. (1992) *Dynamics of Electrophoresis*. VCH, Weinheim.

Rickwood, D. and Hames, B.D. (eds.) (1990) *Gel Electrophoresis of Nucleic Acids, A Practical Approach*, 2nd Edn. IRL Press, Oxford.

Righetti, P.G. (1983) *Isoelectric Focusing: Theory, Methodology and Applications*. Elsevier, Amsterdam.

Righetti, P.G. (1990) *Immobilized pH Gradients: Theory and Methodology*. Elsevier, Amsterdam.

Sambrook, J., Fritsch, E.F. and Maniatis, T. (1989) *Molecular Cloning, A Laboratory Manual*, 2nd Edn. Cold Spring Harbor Laboratory Press, New York.

2 Series and journals

Advances in Electrophoresis (A. Chrambach, M.J. Dunn, and B.J. Radola, eds). VCH, Weinheim.

Analytical Biochemistry. Academic Press, New York.

Electrophoresis. VCH, Weinheim.

INDEX